Smart Systems with Shear Thickening Fluid

Selim Gürgen
Editor

Smart Systems with Shear Thickening Fluid

Springer

Editor
Selim Gürgen
Department of Aeronautical Engineering
Eskişehir Osmangazi University
Eskişehir, Türkiye

ISBN 978-3-031-53569-7 ISBN 978-3-031-53570-3 (eBook)
https://doi.org/10.1007/978-3-031-53570-3

This Springer imprint is published by the registered company Springer Nature Switzerland AG
The registered company address is: Gewerbestrasse 11, 6330 Cham, Switzerland

Paper in this product is recyclable.

Preface

The book *Smart Systems with Shear Thickening Fluid* provides a detailed information about Shear Thickening Fluid (STF)-based smart systems. Engineers, researchers and scientists benefit from this book to understand the smart behavior of STF in different applications. STF exhibits adaptive rheological behavior under different conditions while STF rheology can be tailored for smart applications. Conventional structures or systems gain smart properties with the integration of STF. Some examples are magnetorheological materials, energy harvesting applications, battery electrolytes, vibration damping systems, polishing operations, and protective structures. The content helps understanding the smart concept provided by STF in various applications.

Eskişehir, Türkiye

Selim Gürgen

Contents

About the Editor

Selim Gürgen is an Associate Professor with the Department of Aeronautical Engineering at Eskişehir Osmangazi University, and the Director of Eskişehir Vocational School. He is an Executive Board Member of the ESOGU Aerospace Technologies Research Center. He was a Visiting Researcher at the University of Wollongong for a project on smart materials. He was at the University of Aveiro on natural composites, and worked as a Researcher at the School of Transportation at Anadolu University for 4 years. Before his academic life, he worked as a manufacturing engineer in a joint venture of KMWE-Dutch Aero. Dr. Gürgen has contributed to several industrial and research projects in collaboration with major organizations such as Fiat Automobiles, Turkish Aerospace Industries (TAI), and the Turkish Air Force. His projects were funded by the Scientific and Technological Research Council of Turkey (TÜBİTAK) and the European Union (EU). He has many publications in refereed international journals and conferences. He serves as an editorial member for several international journals. His primary area of research is material science.

Chapter 1
Introduction

Selim Gürgen

Materials science is a constantly evolving and fascinating world, where innovations converge to create solutions for many of the challenges and applications. In recent years, shear thickening fluids (STFs) have emerged as a remarkable class of materials with unique properties and a wide range of practical uses. The concept of shear thickening, or the sudden increase in the viscosity of a fluid under stress, has been the subject of scientific inquiry for centuries. Shear thickening behavior has paved the way for STF integration into smart systems, which are characterized by their ability to adapt and respond intelligently to external stimuli.

In this book, we investigate the various aspects of STFs and their smart applications in-depth, presenting a comprehensive overview of the state-of-the-art in STFs, as well as providing researches and case studies in this field. One of the most interesting STF-based materials is magnetorheological STF. This smart material changes its rheological properties in response to an external magnetic field. When exposed to a magnetic field, magnetorheological STFs undergo rapid changes in viscosity, making them ideal for applications like automotive suspension systems. They allow vehicles to adjust their ride quality and stability in real time, enhancing both comfort and safety [1]. Triboelectric nanogenerators (TENG) is another smart system that is incorporated with STF in recent studies [2, 3]. TENGs are energy-harvesting devices that convert mechanical energy into electricity. By incorporating STFs into TENG designs, these devices can adapt to varying mechanical inputs, increasing their energy output efficiency. This innovation holds potential for self-powered electronics and wearable technology. Energy storage with STF has been a challenging field for recent years. Shear thickening electrolytes in batteries offer a unique solution to the challenges of energy storage. These materials can enhance the safety of lithium-ion batteries by thickening in response to high shear forces, reducing the

S. Gürgen (✉)
Department of Aeronautical Engineering, Eskişehir Osmangazi University, Eskişehir, Türkiye
e-mail: sgurgen@ogu.edu.tr

S. Gürgen (ed.), *Smart Systems with Shear Thickening Fluid*,
https://doi.org/10.1007/978-3-031-53570-3_1

risk of thermal runaway and short-circuits. This results in safer, more reliable energy-storage solutions for various applications, from consumer electronics to electric vehicles [4, 5]. Vibration damping is one of the most suitable fields for benefitting from smart systems. For this reason, STFs have been utilized in vibration damping systems to improve performance and safety in structures and machinery. These materials adapt their viscosity in response to vibrations, effectively dissipating energy and reducing vibrations [6, 7]. This technology benefits industries like defense, aerospace, automotive, construction, and manufacturing, where precise control of vibrations is essential [8]. Surface processing is quite a new field for STF integration. Smart polishing systems employing STFs offer the ability to adjust the material removal rate in real time. Controlling the shear thickening behavior during the process provides more precise and efficient polishing of delicate surfaces as well as reducing material waste and improving product quality [9, 10]. Shear thickening materials have redefined protective gear, as they can remain flexible during regular activities and instantaneously harden upon impact. This innovation provides exceptional protection for individuals in high-risk professions, such as law enforcement and the military, as well as athletes, reducing the risk of injury without compromising mobility [11, 12]. In each of these applications, STFs integrated into smart systems offer a dynamic and adaptive solution to real-world challenges. The ability of these materials to respond intelligently to external forces and stimuli is transforming multiple industries, from transportation and energy storage to manufacturing and personal protection. As research and development continue in this field, researchers can anticipate even more groundbreaking applications and benefits. STFs are poised to be a cornerstone of future innovations in smart systems.

References

1. Zhang, X., Li, W., Gong, X.L.: Study on magnetorheological shear thickening fluid. Smart Mater. Struct. **17**(1), 015051 (2008)
2. Yun, S.Y., Tcho, I.W., Kim, W.G., Kim, D.W., Son, J.H., Lee, S.W., et al.: Mechanically robust triboelectric nanogenerator with a shear thickening fluid for impact monitoring. J. Mater. Chem. A. **10**(19), 10383–10390 (2022)
3. Shang, L., Wu, Z., Li, X., Xu, A., Miao, Y., Xu, W., et al.: A breathable and highly impact-resistant shear-thickened fluid (STF) based TENG via hierarchical liquid-flow spinning for intelligent protection. Nano Energy. **118**, 108955 (2023)
4. Huang, Q., Liang, X., Liu, B., Deng, H.: Research progress of shear thickening electrolyte based on liquid–solid conversion mechanism. Batteries. **9**(7), 384 (2023)
5. Gong, X., Zhang, J., Xuan, S.: Multi-functional systems based on shear thickening fluid. In: Gürgen, S. (ed.) Shear Thickening Fluid [Internet], pp. 53–75. Springer, Cham (2023) Available from: https://link.springer.com/10.1007/978-3-031-25717-9_4 [cited 2023 Oct 27]
6. Sheikhi, M.R., Gürgen, S., Kuşhan, M.C.: Vibration damping systems with shear thickening fluid. In: Gürgen, S. (ed.) Shear Thickening Fluid [Internet], pp. 77–97. Springer, Cham (2023) Available from: https://link.springer.com/10.1007/978-3-031-25717-9_5 [cited 2023 Oct 27]
7. Zhang, X.Z., Li, W.H., Gong, X.L.: The rheology of shear thickening fluid (STF) and the dynamic performance of an STF-filled damper. Smart Mater. Struct. **17**(3), 035027 (2008)

8. Gürgen, S., Sofuoğlu, M.A.: Experimental investigation on vibration characteristics of shear thickening fluid filled CFRP tubes. Compos. Struct. **226**, 111236 (2019 Oct)
9. Lyu, B.H., He, Q.K., Chen, S.H., Shao, Q., Chen, Y., Geng, Z.Y.: Experimental study on shear thickening polishing of cemented carbide insert with complex shape. Int. J. Adv. Manuf. Technol. **103**(1–4), 585–595 (2019)
10. Gürgen, S., Sert, A.: Polishing operation of a steel bar in a shear thickening fluid medium. Compos. Part B Eng. **175**, 107127 (2019)
11. Majumdar, A., Butola, B.S., Srivastava, A.: Development of soft composite materials with improved impact resistance using Kevlar fabric and nano-silica based shear thickening fluid. Mater. Des. **1980–2015**(54), 295–300 (2014)
12. Yay, Ö., Sheikhi, M.R., Kunt, G., Gürgen, S.: Shock absorption in shear-thickening fluid included 3D-printed structures. In: Gürgen, S. (ed.) Shear Thickening Fluids in Protective Applications [Internet], pp. 27–38. Springer, Cham (2024) Available from: https://link.springer.com/10.1007/978-3-031-42951-4_4 [cited 2023 Oct 27]

Chapter 2
Magnetorheological Shear Thickening Fluid

Tongfei Tian

Nomenclature

CBN	Cubic boron nitride
CIP	Carbonyl iron particles
MR	Magnetorheological
MRF	Magnetorheological fluid
MRR	Material removal rate
MRSTF	Magnetorheological shear thickening fluid
MRSTP	Magnetorheological shear thickening polishing
PEG	Polyethylene glycol
STF	Shear thickening fluid

2.1 Introduction

Magnetorheological shear thickening fluid (MRSTF) represents a class of smart materials that have garnered significant attention in recent years due to their unique rheological characteristics and potential applications across various engineering domains. Unlike traditional fluids, MRSTFs exhibit the remarkable ability to undergo rapid changes in viscosity in response to either an applied magnetic field or shear conditions, which was known as shear thickening [1–6]. Unique properties of MRSTF make them particularly intriguing for their potential use in damping systems [7–15] and polishing applications [2, 11, 13, 16–22], such as the innovative magnetorheological shear thickening polishing (MRSTP) process.

T. Tian (✉)
School of Science, Technology and Engineering, University of the Sunshine Coast, Sippy Downs, Australia
e-mail: ttian@usc.edu.au

S. Gürgen (ed.), *Smart Systems with Shear Thickening Fluid*,
https://doi.org/10.1007/978-3-031-53570-3_2

At their core, MRSTFs are suspensions composed of micron-sized magnetic particles dispersed in a shear thickening carrier fluid. The interactions between these magnetic particles, typically influenced by an external magnetic field, lead to the formation of chain structures of iron particles, by which the viscosity of MRSRTFs could be increased [3, 10]. Shear stress could also intricate microstructures within the fluid, causing it to undergo a transition from a liquid-like to a solid-like state [1]. This dynamic behavior holds immense promise for engineering applications where controllable and rapid alterations in material properties are essential [2, 8, 16, 19].

This chapter aims to provide a comprehensive exploration of MRSTFs, delving into their fundamental properties, fabrication of MRSTFs, and diverse applications. It becomes evident that the multifaceted nature of MRSTFs opens avenues for innovations in fields such as damping technology, surface finishing, and multifunctional material design. Through a systematic examination of the current state of knowledge and recent advancements, this chapter seeks to contribute to a deeper understanding of MRSTFs and their potential role in shaping the future of smart materials in engineering.

2.2 MRSTF Properties

Because an MRSTF is a combination of both magnetorheological (MR) fluid and shear thickening fluid (STF), it represents both MR effect and shear thickening effect. In 2008, Zhang et al. firstly mixed MR fluid and STF to produce the MRSTF, which revealed a new direction in smart materials and engineering [12].

MRSTFs exhibit a distinctive MR effect, a property that allows for dynamic control of their rheological characteristics in response to an external magnetic field. The MR effect arises from the presence of magnetizable particles within the fluid, typically composed of materials such as iron, cobalt, or nickel. When subjected to a magnetic field, these particles align and form chain-like structures, altering the overall rheological behavior of the fluid. This alignment process enables engineers to actively modulate the viscosity and stiffness of MRSTFs, providing a mechanism for precise and real-time control in various engineering applications, including dampers and shock absorbers [9]. In 2008, Zhang et al. fabricated MRSTFs and tested them at the magnetic fields of 0, 100, 200, 300, and 400 kA/m. All of the samples showed obvious MR effect. Qian et al. fabricated MRSTF mixed with micro-size cubic boron nitride (CBN) abrasive particles for designing a smart polishing slurry [20]. Figure 2.1 shows the viscosity flow curves of this slurry at various magnetic fields. From the chart, it is obvious that the slurry shows higher viscosities with stronger magnetic fields.

A unique characteristic with an MRSTF is the remarkable shear thickening behavior, wherein the fluid undergoes a transition from a liquid-like to a solid-like state under increasing shear rates. This phenomenon is crucial for applications where sudden changes in material viscosity and stiffness are desirable. The shear thickening properties of MRSTFs are influenced by factors such as critical shear

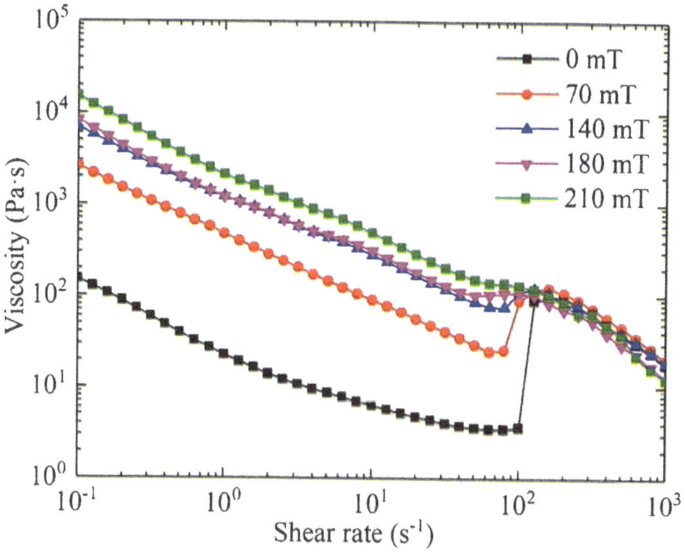

Fig. 2.1 Viscosity flow curves of an MRSTF with CBN particles at various magnetic fields [20]

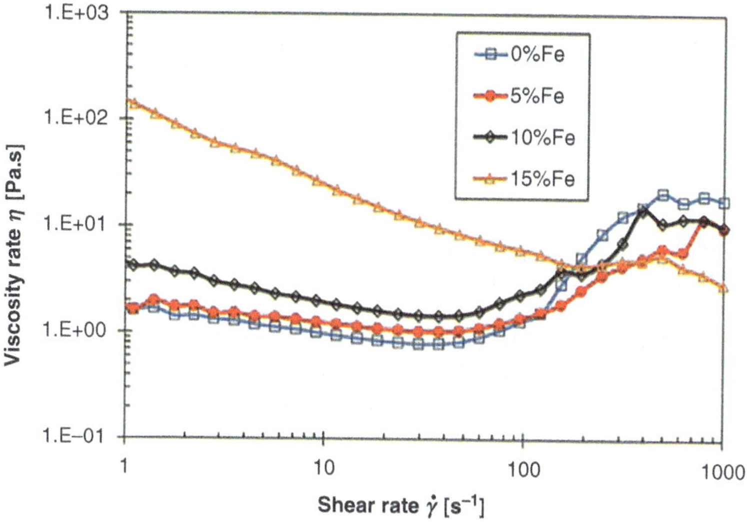

Fig. 2.2 Viscosity flow curves of an MRSTF with different CIP concentrations [3]

rate, particle interactions, and microstructure formation under shear stress. The rheological properties make MRSTFs suitable for applications in polishing processes, and damping systems. The ability to control shear thickening enhances the adaptability of MRSTFs in responding to varying mechanical conditions. Sokolovski et al. fabricated an MRSTF by dispersing different amounts of carbonyl iron particles (CIPs) in an STF [3]. The MRSTF shows a clear shear thickening effect without a magnetic field as shown in Fig. 2.2.

The synergetic effect of MR and shear thickening properties in MRSTFs offer a multifaceted approach for specific engineering requirements. However, the alignment of magnetizable particles under the influence of a magnetic field would diminish the shear thickening response, and thereby providing an additional dimension of control [12]. Sokolovski et al. summarized the behavior of MRSTF with iron particles at different weight concentrations and under different magnetic flux densities [3]. According to the findings in their study, they categorized the MRSTF behavior into three regions, namely STF region, MRSTF region, and MRF region as shown in Fig. 2.3. The limit of their study is that their categorization is only valid for a certain type of MTSTFs, and it could broaden to a wide range of MRSTFs available for various engineering applications.

The dual functionality allows for precise adjustments in material viscosity, stiffness, and damping characteristics, making MRSTFs highly versatile in adaptive structures, shock absorbers, and materials processing. The reversible nature of the MR effect and shear thickening ensures that these changes are dynamic, and can be rapidly adjusted, opening avenues for innovative solutions in fields where dynamic material properties are of paramount importance [3, 8]. Understanding and harnessing this combination of properties are fundamental for optimizing the performance and applications of MRSTFs across diverse engineering disciplines.

Understanding thixotropic behavior in MRSTFs is crucial for applications where controlled variations in viscosity over time are essential. Thixotropy refers to the time-dependent recovery of a material's viscosity after being subjected to shear stress. Zhang et al. found that their MRSTF exhibited a significant thixotropy phenomenon, which reduced the settling problem of MR fluids as well as increasing the

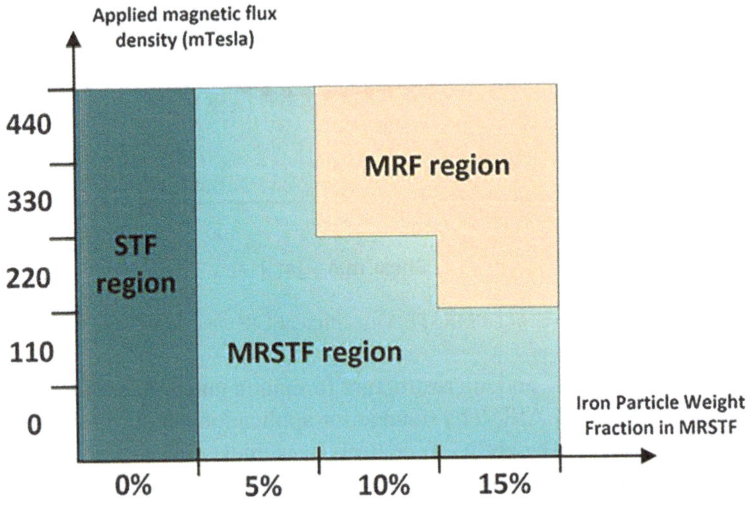

Fig. 2.3 Categorization for a certain type of MRSTF [3]

stability when the weight fraction of STF base was above a threshold value of 15% [5]. The addition of silica particles to the STF base also improved the stability of the iron particle chain structures within the MRSTF in a magnetic field. They also presented experimental results on the steady shear and quasi-static shear behavior of the MRSTF, as well as theoretical modeling to verify the experimental studies. Qian et al. stated that thixotropic behavior of MRSTF was evident in rheological experiments [20]. Their MRSTF exhibited thixotropic properties, which were beneficial in preventing the settlement of particles. Additionally, the study revealed that thixotropic properties of MRSTF were influenced by concentration of CIPs, and strength of magnetic field. From these findings, thixotropy in MRSTF could be controlled, and optimized by adjusting CIP concentration and magnetic field strength, which were valuable for practical applications as such in polishing processes. Zhang et al. found that higher concentrations of fumed silica in MRSTF provided a thixotropic medium for stabilizing the dispersion of magnetic particles, and thereby preventing sedimentation [12]. Their MRSTF exhibited a magnetic-field-controllable characteristic and reversible behavior, with a drastic jump in complex viscosity beyond a critical dynamic shear rate.

2.3 MRSTF Fabrication

MRSTFs consist of two main components; (i) magnetizable particles and (ii) STF as the carrier liquid. The mixture of these components allows MRSTFs to exhibit unique MR behavior and shear thickening effect.

Magnetizable particles, typically micron-sized, are dispersed in the carrier liquid. These particles are composed of materials with high magnetic permeability, such as iron, cobalt, nickel, or iron oxides like magnetite. Commonly used magnetic particles in MRSTFs are CIPs, mostly in micron-size. C3518 from Sigma-Aldrich, with 5 μm mean particle size, is one of the most preferred CIPs in this field [1, 3–5, 8, 10, 12, 14]. Ming et al. used CIPs with diameters of 3–5 μm to synthesis their MRSTF employed in surface finishing operations [19]. Du et al. used 3.5 μm-sized CIPs to fabricate a multifunctional STF, which was prepared for a shock absorber [7]. Another MRSTF was fabricated by distributing 3 μm-sized CIPs in an STF base by Yang et al. [15]

Carrier liquid serves as a base medium for MRSTF, providing a liquid matrix in which magnetizable particles are suspended. Carrier liquids in MRSTFs should be an STF to ensure the shear thickening effect in the material. STFs also have their own carrier fluids and nanoparticles, where fumed silica with ethylene glycol is a popular combination of STF. The fumed silica S5505 from Sigma-Aldrich is one of the mostly used nanoparticles in STF, which has a primary particle size of 14 nm and a specific surface area of 200 m^2/g [1, 3, 5, 8, 10, 12, 14]. Ming et al. used 7–40 nm-sized silica particles with the surface area of 300 m^2/g in their STF [18, 19]. Starch polymer particles ((C6H10O5)n) were also used as the solid phase for fabricating an STF in another study [21]. A common carrier liquid for STFs is ethylene glycol with a density of 1.113 g/ml (102,466, Sigma-Aldrich) [1, 3, 5, 8, 10, 12, 14]. Polyethylene glycol (PEG) is another carrier liquid widely used for synthesizing STFs [6, 11, 15, 20, 21]. The weight fraction of nanoparticles in STFs is generally in a range from 20 to 60% [3, 5, 8, 20].

The weight fraction of CIP in MRSTFs is significant as it directly affects the rheological properties and magnetizing performance of the material. Zhang et al. investigated an MRSTF by varying the loading of iron particles in MRSTF [12]. Their study showed that lower weight fractions of iron particles allowed the observation of both the MR effect and shear thickening behavior, while higher concentrations lead to a strong MR effect by limiting the shear thickening behavior. Their sample with the CIP:STF ratio of 200:100 only showed the MR effect without shear thickening. Sokolovski et al. mixed CIPs with STF at different concentrations [3]. They found that the weight fraction of CIPs significantly affected the critical shear rate and the viscosity profile of MRSTF under magnetic fluxes.

In some special applications, additives can be included in the MRSTF such as abrasive particle additives in MRSTF for designing MRSTP. Ren et al. added abrasive alumina particles to an MRSTF for increasing the material removal performance in surface polishing processes, where the MRSTF was a promising candidate as a polishing slurry [21]. Micro-sized cBN particles were also used to enhance the polishing performance of MRSTP slurries. The addition of CBN particles aimed to improve the material removal rate (MRR) and polishing efficiency, contributing to the development of a high-efficiency and ultra-precision polishing process [20].

Fabrication process of MRSTF has two main steps. The first step is the preparation of STF, which is the carrier liquid of MRSTF. The second step is the addition of magnetizable particles into the STF. The magnetizable particles are thoroughly mixed with the STF under high shear conditions, and the resulting mixture is then degassed in a vacuum chamber. The weight fractions of solid particles in STF and magnetizable particles in MRSTF can be varied for optimum properties in the applications. Hence, a wide range of MRSTFs can be fabricated depending on the

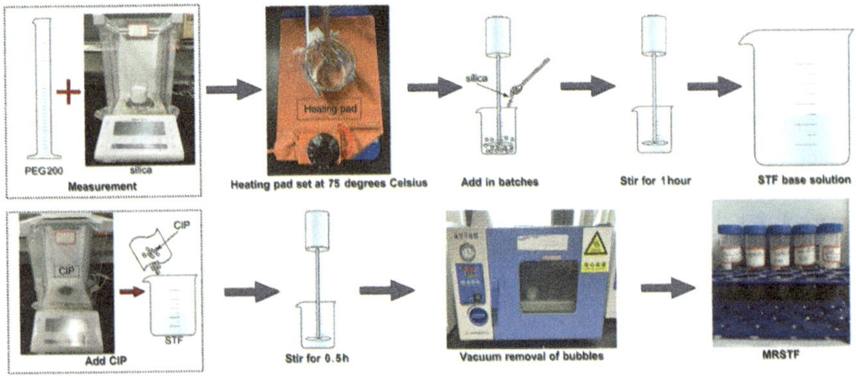

Fig. 2.4 MRSTF fabrication process [6]

requirements in the applications. Figure 2.4 shows a schematic illustration of MRSTF fabrication process according to Zhao et al. [6]

2.4 MRSTF Modeling

In the field of MRSTF, the development of effective constitutive models plays a pivotal role in understanding and optimizing the rheological properties for specific applications. This section explores several noteworthy models proposed by various researchers, shedding light on the diverse approaches employed to comprehend the intricate behavior of MRSTFs. From phenomenological theories to viscoelastic models, each model discussed herein contributes to the comprehensive understanding of MRSTF characteristics, paving the way for advancements in controllable damping equipment and material removal processes.

Zhao et al. proposed a phenomenological theory-based constitutive model for MRSTF called the M-S model [6]. The model was derived based on the geometric transformation of shear stress curves by considering the inhibitory relationship between the MR effect and shear thickening behavior. The model provided a correlation between materials composition and rheological behavior, which helps fabricating the MRSTFs with required MR performances. The model was validated by fitting curves of shear stress, and it was found that the deviations in the predicted values are in an acceptable range. The M-S model contributed to the understanding of the MR behavior in MRSTFs and offered an effective guidance for future studies and engineering applications mainly in the field of controllable damping systems. Peng et al. proposed a four-parameter viscoelastic model to predict the mechanical response of the MRSTFs at various conditions [8]. The MATLAB optimization algorithm was used to identify the model parameters, and the comparison of experimental data with the model prediction results showed that the four-parameter model could accurately predict the viscoelastic properties of MRSTFs. Yang et al. established theoretical models to describe the frequency shift performance of an MRSTF based damper [15]. Two damping models were created, with one based on energy storage stiffness and the other on energy dissipation stiffness. It was observed that the nonlinear damping model more effectively reproduced the nonlinear stiffness, nonlinear damping, and shear rate-related characteristics of the MRSTF damper. Ren et al. presented a model based on the Bingham model to express the constitutive relation of the MR fluid [21]. The model demonstrated the relationship between the viscosity and shear stress of the MRSTF under different conditions, such as magnetic field intensity and iron particle concentration.

When MRSTF was used in MRSTP, there were some models of the material removal proposed. Ma et al. proposed another model of MRR of the MRSTP process, which was established based on the characteristics of polishing pressure [2]. It was experimentally validated, showing an average error between theoretical and experimental results of 4.1%. Sun et al. established a mathematical model of the material removal relating the MRR to the composition of the MRSTF media, including the volume fractions of CIPs and STF [13]. The model showed that the MRR was inversely proportional to the volume fraction of the STF, while it was directly proportional to the volume fraction of CIPs. Li et al. presented a model of MRR for the MSTP process, which was established based upon the calculations of the indentation depth and indentation diameter formed by abrasive particles [17]. It predicted the feasibility of the MRSTP method by demonstrating the initial and machined surfaces using Gaussian filtering. Ming et al. established a numerical prediction model of surface roughness for MRSTP, which was based on both impact energy and material removal models, and the numerical boundaries of polishing flow velocity and shear stress were obtained through finite element calculation [18]. Qian et al. presented a mathematical model to define the variation of shear rate related to the shear stress in MRSTFs [20]. The model was based on the Herschel-Bulkley model and took into account the effects of magnetic flux density and particle size on the rheological behavior of the MRSTFs. Tian et al. introduced a material removal model for MRSTP that integrated principles from magnetohydrodynamics, non-Newtonian fluid kinematics, and microscopic contact mechanics [22]. This model accurately forecasted the material removal depth of the workpiece under various processing parameters while achieving an average prediction error of under 5%. Their model provided a theoretical guidance for the practical applications of MRSTP. This model demonstrated the potential of MRSTP in the field of ultra-precision machining of difficult-to-machine materials, such as Ti-6Al-4V alloy.

In summary, the presented models offer a rich tapestry of insights into the complex world of MRSTFs. The proposed models collectively contribute to the evolving landscape of MRSTF research, offering a foundation for both theoretical understanding and practical applications in controllable damping and material removal processes.

2.5 MRSTF in Engineering Applications

MRSTFs have drawn significant attention due to their unique rheological properties, offering versatile applications across various engineering domains. This section provides an overview of key applications, where MRSTFs have demonstrated exceptional performance.

2.5.1 Damping Systems and Shock Absorbers

MRSTFs find extensive use in damping systems and shock absorbers, where the ability to dynamically adjust viscosity and stiffness in response to varying mechanical conditions is crucial. The shear thickening behavior, coupled with the MR effect, allows for precise control of damping characteristics, resulting in improved vehicle suspension systems, impact mitigation, and vibration control. Du et al. integrated an MRSTF into a shock absorber and found that the rebound force of the absorber was significantly higher than that of the MR fluid by applying current to the damping system [7]. From this study, MRSTF integration into the damping devices demonstrated the potential of this smart fluid in improving the performance of shock absorbers. As shown in Fig. 2.5, Liu et al. used an MRSTF in a shock absorber, and compared its damping force and variation rules with a commercial MR fluid at different currents and velocities [11]. The results showed that the incremental rate of the damping force in the MRSTF-based shock absorber was higher than that in the MR fluid-based absorber, indicating that the energy dissipation performance of MRSTF, having distributed magnetic particles in an STF medium, increased quicker than that of MR fluid due to the shear thickening rheology in MRSTF. Yang et al. fabricated an MRSTF with different weight fractions of iron particles, and then filled the MRSTF into a bypass linear damper to investigate the effect of iron particle loading [14]. It was stated that shear thickening effect in the MRSTF-based damper faded away by increasing the iron particle concentration in the material. However, MR effect got stronger by including more iron particles in this smart fluid. Furthermore, the effective stiffness of the MRSTF-based damper was almost independent from the variable frequency, which means that the damping system was not bound to the excitation frequency. Yang et al. prepared MRSTF samples based on fumed silica, polyethylene glycol, and different amounts of iron powder [15]. They

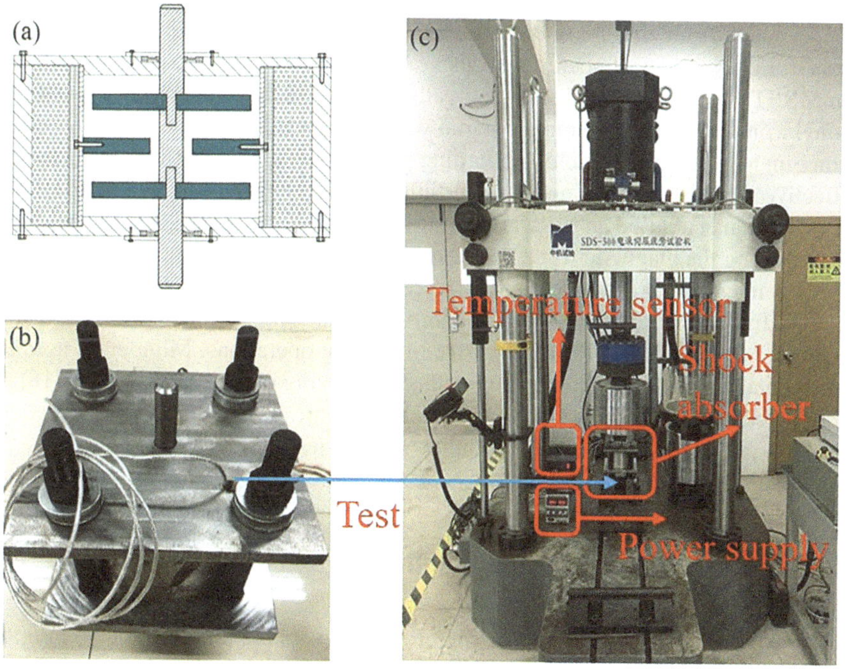

Fig. 2.5 MRSTF-based shock absorber and testing system [11]

analyzed the rheological properties of the MRSTFs, and the results showed that the damping force in the MRSTF-based damper varies with the strength of magnetic flux, vibration frequency, and vibration amplitude. The damper could provide enhanced vibration suppression performance by controlling the shear rate in the piston gap. Moreover, it was possible to actively control the damping force by tuning the magnetic field through the damping coil.

2.5.2 *Polishing and Surface Finishing Processes*

In materials processing, MRSTFs have been employed in polishing and surface finishing applications. The shear thickening properties aid in achieving controlled abrasive action, providing enhanced material removal rates. Moreover, the adaptability of MRSTFs to magnetic fields allows for fine-tuning surface finishing processes, contributing to the advancement of precision manufacturing techniques. Ma et al. integrated an MRSTF into a newly designed MRSTP apparatus that formed magnetic force reinforced flexible abrasive clusters under a magnetic field [2]. The clusters, containing additives, silica, abrasives, and carbonyl iron powder, were employed to have material removal by the relative motion generated with the workpiece. Ming et al. developed an MRSTF for surface finishing operations by considering the chain-forming effect of magnetic particle polarization on the rheological properties [19]. The MRSTF was used in spherical polishing of refractory workpieces made from zirconium dioxide, demonstrating the potential of MRSTP process from the aspects of operational costs, and process efficiency. Sun et al. prepared an MRSTF by using fumed silica, polyethylene glycol, abrasive particles, and carbonyl iron particles. This smart material was developed for surface finishing operations under a magnetic field. According to this work, the MRSTF provided an effective material removal, and surface-finishing mechanism for manufacturing sector [13]. Fan et al. used an MRSTF in a surface-finishing process [16]. The MRSTF, drawn along the magnetic force lines like a pliable brush due to the influence of the applied magnetic pole, successfully decreased the surface roughness of the test sample from an initial value of 1.17 mm to 54 nm. The test results illustrated the significant efficacy of MRSTF in surface-finishing operations. Ming et al. used an MRSTF as a polishing slurry in the surface processing of complex surfaces [18]. It was stated that MRSTF is a high-performance smart material, providing excellent material removal mechanism for complex geometries.

2.6 Future Directions and Challenges

In the MRSTF applications, future research endeavors are poised to delve into the refinement of advanced material formulations. This entails a meticulous exploration of novel combinations involving base fluids, magnetic particles, and additives, aimed at tailoring the rheological properties of MRSTFs to achieve heightened performance in specific engineering applications. The evolution of MRSTFs with multifunctional capabilities is also a promising trajectory, encompassing integration of functionalities like self-healing, shape memory, or enhanced thermal conductivity. As smart manufacturing principles take center stage in Industry 4.0, there is a growing focus on incorporating MRSTFs into adaptive manufacturing systems, allowing for real-time adjustments and improvements in processes, such as polishing, and surface finishing. Additionally, the exploration of biomedical applications, including medical devices and drug delivery systems, showcases the potential versatility of MRSTFs in advancing materials science.

Navigating the future of MRSTFs is not without its share of challenges. Stability and agglomeration issues concerning magnetic particles within these fluids require meticulous attention, as achieving long-term stability while mitigating particle clustering remains a persistent challenge. The temperature sensitivity exhibited by many MRSTFs poses another hurdle, necessitating the development of formulations capable of maintaining consistent performance across varying temperature ranges. Scaling up production for industrial applications introduces challenges related to ensuring consistent quality and cost-effectiveness, demanding innovative manufacturing processes and rigorous quality control measures. In the context of potential biomedical applications, the biocompatibility and toxicity of MRSTFs emerge as critical considerations, prompting comprehensive studies on their interaction with biological systems for safe implementation in medical contexts. The modeling and simulation complexity associated with predicting the behavior of MRSTFs under diverse conditions further underscores the need for ongoing research to refine and validate models, facilitating improved predictive capabilities and aiding in the design and optimization of applications. Addressing these challenges will be pivotal in unlocking the full potential of MRSTFs, and fostering advancements in materials science and engineering.

2.7 Conclusions

In conclusion, the comprehensive exploration into MRSTFs has illuminated their unique rheological characteristics and diverse applications across engineering domains. The inherent ability of MRSTFs to dynamically adjust viscosity and stiffness in response to external stimuli, such as magnetic fields, opens avenues for innovative solutions in damping systems, shock absorbers, and polishing applications, notably in the MRSTP process. The properties of MRSTFs, including the MR effect and shear thickening behavior, provide a multifaceted approach to tailoring their behavior for specific engineering requirements. The synergy of MR and shear thickening properties allows for precise adjustments in material properties, offering versatility in adaptive structures, shock absorbers, and materials processing. Thixotropic behavior further enhances stability, addressing concerns related to settling and sedimentation.

The fabrication of MRSTFs includes micron-sized magnetic particles, such as CIPs dispersed in a shear thickening carrier liquid. The weight fraction of CIPs and the choice of carrier liquid significantly affect the rheological characteristics of MRSTFs. Additives, like abrasive particles, contribute to enhancing the polishing performance in MRSTP operations.

Various constitutive models have been proposed to understand and optimize the rheological properties of MRSTFs, offering insights for controllable damping equipment and material removal processes. Applications of MRSTFs extend to damping systems, shock absorbers, and polishing processes, where their unique properties lead to improved performance and efficiency.

Looking ahead, future research directions involve refining material formulations, exploring multifunctional capabilities, and integrating MRSTFs into adaptive manufacturing systems and shock-absorbing devices. Challenges, including stability issues, temperature sensitivity, and scalability for industrial applications, underscore the need for ongoing research and innovative solutions. The potential biomedical applications of MRSTFs also require thorough investigation to ensure biocompatibility and safety. In navigating the future of MRSTFs, addressing these challenges will be pivotal in unlocking their full potential and fostering advancements in materials science and engineering.

References

1. Li, W.H., Zhang, X.Z.: Rheology of magnetorheological shear thickening fluids. Adv. Mater. Res. **32**, 161–164 (2008). https://doi.org/10.1088/1361-665X/ac7bbd
2. Ma, Z., Tian, Y., Qian, C., Ahmad, S., Fan, Z., Sun, Z.: Modeling and simulation of material removal characteristics in magnetorheological shear thickening polishing. Int. J. Adv. Manuf. Technol. **128**(5), 2319–2331 (2023). https://doi.org/10.1007/s00170-023-12093-7
3. Sokolovski, V., Tian, T., Ding, J., Li, W.: Fabrication and characterisation of magnetorheological shear thickening fluids. Front. Mater. Sci. **7**, 7 (2020). https://doi.org/10.3389/fmats.2020.595100
4. Yu, G.-J., Zhu, S.-J., Du, C.-B., Wang, L.-Y., Huang, J.-C.: Study on the magnetic rate sensitivity and shear properties of magnetorheological shear thickening fluid. J Magn. **28**(3), 290–298 (2023). https://doi.org/10.4283/JMAG.2023.28.3.290
5. Zhang, X., Li, W., Gong, X.: Thixotropy of MR shear-thickening fluids. Smart Mater. Struct. **19**(12), 125012 (2010). https://doi.org/10.1088/0964-1726/19/12/125012
6. Zhao, Q., Wang, K., Yuan, J., Jiang, H.: Research on rheological properties and phenomenological theory-based constitutive model of magnetorheological shear thickening fluids. Smart Mater. Struct. **32**, 32 (2023). https://doi.org/10.1088/1361-665X/acf423
7. Du, C., Zeng, F., Liu, B.: The preparation of multifunctional shear thickening fluid and the application in shock absorber. IOP Conf. Ser. Earth Environ. Sci. **639**(1), 012006 (2021). https://doi.org/10.1088/1755-1315/639/1/012006
8. Peng, G., Li, W., Tian, T., Ding, J., Nakano, M.: Experimental and modeling study of viscoelastic behaviors of magneto-rheological shear thickening fluids. Korea-Aust. Rheol. J. **26**, 149–158 (2014). https://doi.org/10.1007/s13367-014-0015-3
9. Hu, G., Ying, S., Qi, H., Yu, L., Li, G.: Design, analysis and optimization of a hybrid fluid flow magnetorheological damper based on multiphysics coupling model. Mech. Syst. Signal Process. **205**, 110877 (2023). https://doi.org/10.1016/j.ymssp.2023.110877

10. Li, W., Nakano, M., Tian, T., Totsuka, A., Sato, C.: Viscoelastic properties of MR shear thickening fluids. J. Fluid Sci. Technol. **9**(2), JFST0019-JFST (2014). https://doi.org/10.1299/jfst.2014jfst0019
11. Liu, B., Du, C., Deng, H., Fan, Z., Zhang, J., Zeng, F., Fu, Y., Gong, X.: Mechanical properties of magneto-sensitive shear thickening fluid absorber and application potential in a vehicle. Compos. Part A Appl. Sci. Manuf. **154**, 106782 (2022). https://doi.org/10.1016/j.compositesa.2021.106782
12. Zhang, X., Li, W., Gong, X.L.: Study on magnetorheological shear thickening fluid. Smart Mater. Struct. **17**(1), 015051 (2008). https://doi.org/10.1088/0964-1726/17/1/015051
13. Sun, Z., Fan, Z., Tian, Y., Qian, C., Ma, Z.: Investigation on magnetorheological shear thickening finishing (MSTF) with radially slotted magnetic pole for free-form surface. Int. J. Adv. Manuf. Technol. **123**(9), 3313–3327 (2022). https://doi.org/10.1007/s00170-022-10313-0
14. Yang, J., Sun, S., Li, W., Du, H., Alici, G., Nakano, M.: Development of a linear damper working with magnetorheological shear thickening fluids. J. Intell. Mater. Syst. Struct. **26**(14), 1811–1817 (2015). https://doi.org/10.1177/1045389x15577653
15. Yang, S., Huang, X., Yang, M., Li, W., Qing, Z.: Rheological behavior of a magnetorheological shear-thickening fluid (MRSTF) and dynamic characteristics of an MRSTF damper. 2022. https://doi.org/10.21203/rs.3.rs-1363595/v1
16. Fan, Z., Tian, Y., Zhou, Q., Shi, C.: A magnetic shear thickening media in magnetic field–assisted surface finishing. Proc. Inst. Mech. Eng. B J. Eng. Manuf. **234**(6–7), 1069–1072 (2020). https://doi.org/10.1177/0954405419896119
17. Li, J., Fan, Z., Yang, Z., Tian, Y., Gao, J.: Simulation and modeling of magnetorheological shear thickening polishing processes for slender tube. J. Mater. Res. **25**, 480–496 (2023). https://doi.org/10.1016/j.jmrt.2023.05.226
18. Ming, Y., Huang, X., Cai, Y., Zhou, D.: Research on prediction method of surface roughness in weak magnetorheological shear thickening fluid polishing. Int. J. Adv. Manuf. Technol. **124**(7), 2659–2673 (2023). https://doi.org/10.1007/s00170-022-10668-4
19. Ming, Y., Huang, X.M., Zhou, D.D., Zeng, Q., Li, H.Y.: Rheological properties of magnetic field-assisted thickening fluid and high-efficiency spherical polishing of ZrO2 ceramics. Int. J. Adv. Manuf. Technol. **121**(1), 1049–1061 (2022). https://doi.org/10.1007/s00170-022-09344-4
20. Qian, C., Tian, Y., Fan, Z., Sun, Z., Ma, Z.: Investigation on rheological characteristics of magnetorheological shear thickening fluids mixed with micro CBN abrasive particles. Smart Mater. Struct. **31**(9), 095004 (2022). https://doi.org/10.1088/1361-665X/ac7bbd
21. Ren, Y., Yang, S., Huang, X., Ming, Y., Li, W.: Research on the rheological characteristic of magnetorheological shear thickening fluid for polishing process. Int. J. Adv. Manuf. Technol. **117**(1), 413–423 (2021). https://doi.org/10.1007/s00170-021-07706-y
22. Tian, Y., Ma, Z., Ahmad, S., Qian, C., Ma, X., Yuan, X., Fan, Z.: Theoretical and experimental investigation of material removal rate in magnetorheological shear thickening polishing of Ti-6Al-4V alloy. J. Manuf. Sci. Eng. **146**, 1–46 (2023). https://doi.org/10.1115/1.4063984

Chapter 3
Shear Thickening Fluid in Triboelectric Nanogenerators

Mahdi Hasanzadeh and Selim Gürgen

3.1 Introduction

Smart wearable electronics have achieved revolutionary progress and have shown wide applications in sensors, artificial skins, touch panels, and smart textiles. However, the rigidity, heaviness, and bulky volume of the external power sources hindered their practical applications as wearable electronics. In this regard, developing novel self-powered electronics harvesting energy and transforming it into electric energy is highly suggested by researchers [1, 2]. Different forms of energy sources, such as mechanical energy, thermal energy, chemical energy, solar energy, and biological energy, can be collected and used in the surrounding environment. Among them, mechanical energy is the most ubiquitously distributed energy source. However, the vast majority of this energy is wasted, through walking, machine vibrations, wind, rain, and ocean waves [3].

In the past decade, different types of nanogenerators (NGs) based on the effect of piezoelectricity and triboelectricity have been developed primarily by scientists to effectively convert mechanical energy into electricity [2]. Triboelectric nanogenerators (TENGs) have exhibited favorable advantages in wearable electronics to harvest mechanical energy for electronic applications requiring low-power. The triboelectric effect was regarded as an adverse effect in industry, because it induced the electrostatic charges that may lead to electronic damage, dielectric breakdown, dust explosions, ignition, etc. However, from an energy aspect, through a

M. Hasanzadeh (✉)
Department of Textile Engineering, Yazd University, Yazd, Iran

Department of Aeronautical Engineering, Eskişehir Osmangazi University, Eskişehir, Türkiye
e-mail: m.hasanzadeh@yazd.ac.ir

S. Gürgen
Department of Aeronautical Engineering, Eskişehir Osmangazi University, Eskişehir, Türkiye

S. Gürgen (ed.), *Smart Systems with Shear Thickening Fluid*,
https://doi.org/10.1007/978-3-031-53570-3_3

conjunction of triboelectric and electrostatic effects, TENGs can convert mechanical energy, which is usually wasted in our everyday lives, into electric outputs. Particularly, due to their high efficiency and flexibility, TENGs can be attached to the human body and used as portable power sources for human electronic devices. The TENGs have widely been demonstrated in various potential applications, such as healthcare, sports, communication, security, renewable energy, and personal electronics [4]. The ability of TENGs to operate in the low-frequency regimes, which is inherent to human motions, makes them particularly interesting for wearable applications.

3.2 Working Modes and the Principle of TENG

Generally, the TENGs operate through the contact of materials, followed by triboelectrification and charge induction. During the past decade, TENGs with different architecture and designs have been developed by researchers. They can be classified according to four different modes of operation. Based on the various types of driving mechanical energy, the operation principles of TENGs were divided into four modes. Figure 3.1 shows the four working modes of TENGs, including the modes (i) vertical contact-separation, (ii) lateral sliding, (iii) single electrode, and (iv)

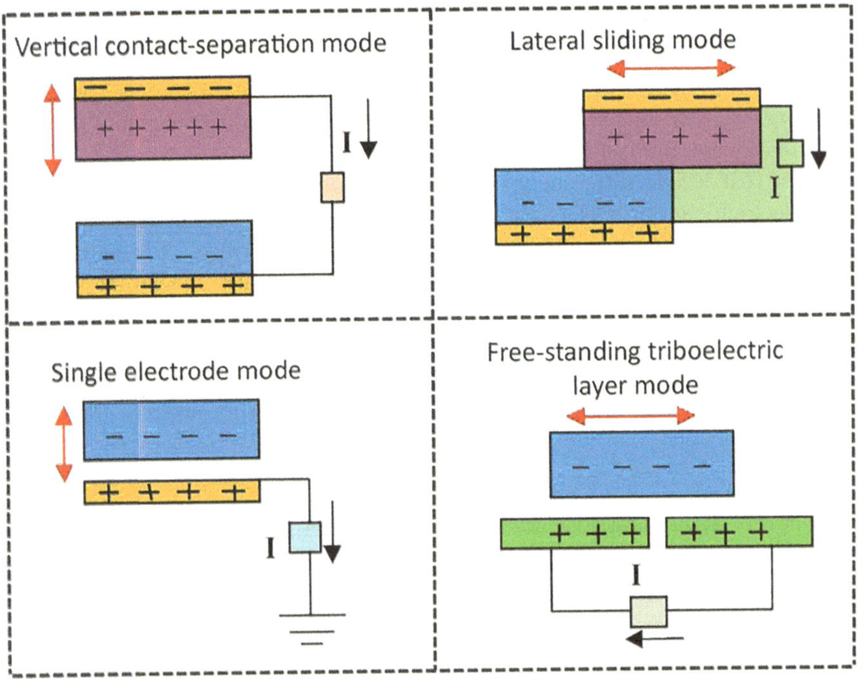

Fig. 3.1 Four fundamental working modes of TENGs

free-standing triboelectric layer [5]. The first TENG, which was developed and investigated in 2012 was the vertical contact-separation mode. Consequently, it encompasses the majority of publications on flexible TENGs. Later, in 2013, lateral sliding mode TENGs were developed. Since then, several studies and experiments have focused on these systems, improving the TENG performance. However, there have only been a few examples of flexible TENGs. Similar to this, single-electrode mode TENGs debuted in 2013. Numerous studies and adaptable applications have been published as a result of the new integration opportunities they present. A most recent study was conducted on free-standing triboelectric layer mode TENGs, which were first proposed in 2014. Although this structure is relatively new, its performance and integration potential are very high. All of these various operating modes complement the others, enabling flexible TENGs to be effectively integrated and used in different applications requiring mechanical energy harvesters and/or self-powered sensors [6].

Figure 3.2 shows the detailed working principle of TENGs in short-circuit conditions. The two tribomaterials come into contact when the TENG is pressed. There will be net negative charges on the surface of the layer to attract electrons and net positive charges on the other. This is due to the different ways that different materials can attract electrons. The interfacial tribo-charges in the two tribo-materials will be separated as they separate, creating an electrical potential that will cause electrons in the attached electrodes to be driven to flow from one side to the other. Electrons continue to flow during this process up until the TENG is completely released, as shown in Fig. 3.2 (state V). The amount of transferred charges and the induced potential difference have now both reached their maximum values. The potential difference will decrease as the two tribo-materials get closer, and the electrons will start to flow back (state VI). As a result, the entire procedure will produce an output of alternate current (AC) pulses.

Generally, it is possible to construct a TENG using any material that exhibits a distinct charge affinity. This leads to different kinds of materials, located at opposite ends of the triboelectric series, achieving high performance. Notably, silicon, polytetrafluoroethylene (PTFE), and polydimethylsiloxane (PDMS) are extensively

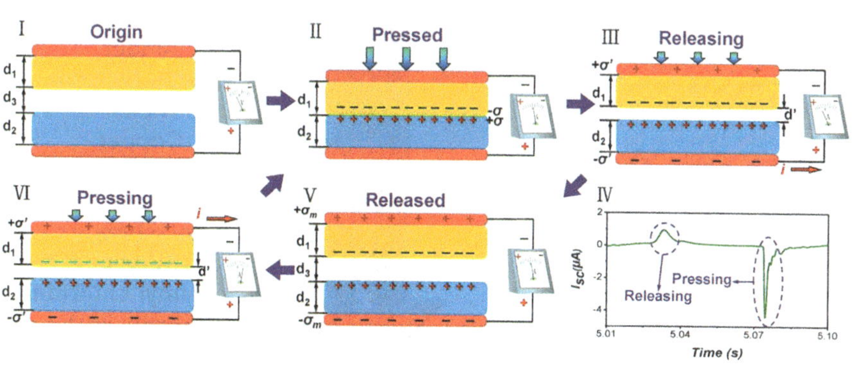

Fig. 3.2 Working principle of TENGs in short-circuit condition [7]

utilized materials for achieving net negative triboelectric charges, while most of the textile materials, such as nylon, cotton, and silk, as well as metals such as aluminum, copper, nickel, and silver are frequently used as the positive triboelectric materials [8]. Figure 3.3 shows a series of triboelectric materials. Currently, the majority of TENGs are made from flexible polymers, making them highly portable, cost-effective, and easy to fabricate. However, TENGs designed for high-temperature applications are constructed from ceramic materials to withstand extreme operating conditions.

3.3 Textile-Based TENGs

Textiles are used in a variety of industries in addition to clothing, such as home furnishings, outdoor gear, transportation, engineering, and medical applications. Textile-based TENGs exhibit wearable, flexible, breathable, comfortable, and mechanically robust advantages, making them suitable for designing wearable devices. Zhong et al. [9] introduced the textile-based TENG, which consisted of carbon nanotubes, cotton threads, and polytetrafluoroethylene (PTFE) aqueous suspension for the first time. Textile-based TENG was capable of converting mechanical energy from human motion into electricity having a power density of 0.1 $\mu W\ cm^{-2}$. Many textile-based TENGs have been created thus far in an effort to

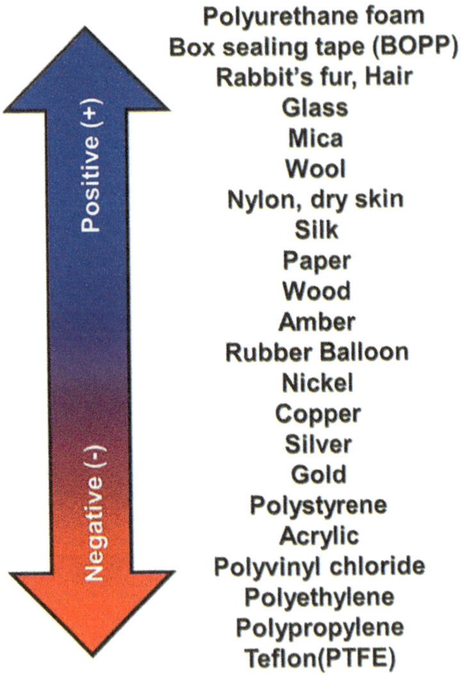

Fig. 3.3 A conventional triboelectric series of materials [8]

further enhance the wearability and comfort properties of TENGs. For the construction of wearable TENG devices, textile materials are the best materials due to their benefits of softness, affordability, and comfort when worn. The possibility of self-powered tactile sensors for human-machine interfaces and wireless music controllers was also presented, along with elastic fibers with the ability to harvest energy from body motion. The textile-based TENGs can be classified into two categories, namely fabric-based TENG and the fiber-based TENG [10]. The fabric-based TENG is the prevailing type of textile-based TENG, which utilizes fabric as a substrate or triboelectric material. The fiber-based TENG is made of single or multiple fibers that can be directly employed as a TENG or be woven or knitted to create a fabric.

According to the literature, a portable TENG composed of nylon and Dacron fabric has demonstrated the ability to effectively harvest energy from human motion and generated up to 2 kV voltages. Cotton-based TENGs have the potential to capture direct sensory information about the human body such as contact force, sweat levels, and physiological motions. The incorporation of polyacrylonitrile (PAN) into TENG has endowed the device with machine-washable and breathable properties. Additionally, since miniaturization is better for wearable technology, all-in-one textile-based TENGs have been prepared benefitting from the fibers. The polyester fiber-based TENG not only demonstrated effective energy harvesting but also demonstrated effective bacterial disinfection. A 3D honeycomb-structured TENG, made with flame-retardant yarn, has the potential to function as a self-powered escape device in fire rescue situations. These findings suggest that functional textiles will give TENGs new qualities that will increase the range of their practical applications. However, it should be noted that currently reported TENG only harvests slight mechanical energy, including wind blowing, water dropping, and body motion. Moreover, the weak mechanical performance of conventional textile-based TENGs makes them susceptible to damage from severe impact loads, which can lead to the collapse of the structure and a decline in electrical performance. Hence, it is necessary to develop a new textile-based TENG with improved anti-impact properties so that it can be used for smart protective systems.

3.4 Shear Thickening Materials

Recently, a successful strategy has been implemented to enhance the mechanical properties of high-performance textiles, such as Kevlar and UHMWPE, by introducing polymer composites onto the fabric. In particular, it has been demonstrated that the incorporation of shear thickening materials, possessing rate-dependent mechanical properties, can significantly improve the frictional interactions between fibers and anti-impact behavior of composites [11, 12]. Generally, shear thickening materials consist of shear thickening fluid (STF), and shear stiffening gel (SSG). STF exhibits a dramatic increase in viscosity when the shear rate surpasses a critical value, which promptly recovers upon unloading the applied shear rate. STF is composed of a liquid medium, such as water or oil that contains solid particles [13, 14]. The defining characteristic of STFs is the interactions between these particles, which leads to the observed phenomenon of shear thickening. When subjected to low shear stress conditions, the particles within the fluid are relatively distant from each other and move independently, allowing the fluid to flow with ease. However, when subjected to high shear rates, the particles begin to interact and form temporary networks or clusters. These particle networks impede the flow of the fluid, resulting in a significant increase in viscosity. As the shear stress increases, the particle interactions become stronger, leading to greater resistance to flow. This phenomenon is known as shear thickening, which provides enhanced protection against impact and deformation [15–18]. Figure 3.4 illustrates the shear thickening behavior under escalating shear stress or shear rates.

STF has been widely utilized in various applications, including protection and damping, as it can dissipate impact energy. By integrating STFs into fabrics or coatings, it is feasible to fabricate materials that exhibit flexibility and comfort under typical circumstances but stiffen upon impact. This characteristic endows them with the ability to absorb and distribute the force of impacts, thereby providing superior protection. STFs also hold significant potential for deployment in other domains, such as robotics, automotive engineering, and manufacturing. They can be employed in damping systems, where their augmented viscosity under impact facilitates the absorption of vibrations and shocks. Furthermore, STFs can be utilized in advanced lubricants, drilling fluids, and even in the development of flexible electronics. Significant interest has been shown in the utilization of shear thickening materials in body armor. With regard to improving ballistic, stab, and puncture resistance without significantly increasing the thickness or stiffness of the fabrics, they have specifically been combined with high-performance fabrics. Numerous research reports in the literature have focused on improving the impact-resistance performance of high-performance textiles with STF and SSG. Early advancements in this area were found in developing soft body armor using STF in conjunction with Kevlar fabric [19]. The experimental findings demonstrate that Kevlar fabric can be made more resistant to ballistic penetration by impregnating it with STF based on

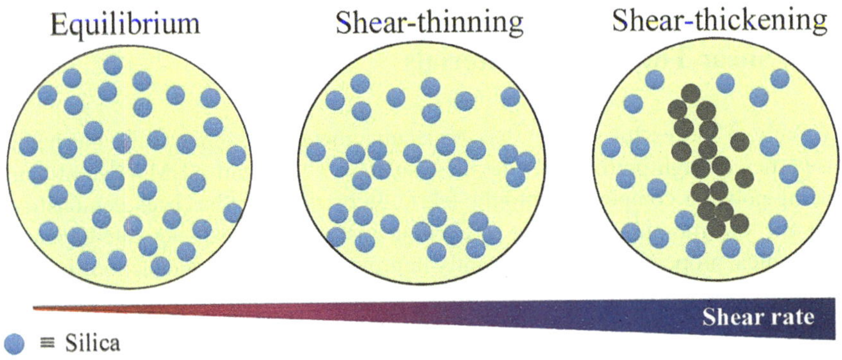

Fig. 3.4 Schematic representation of reversible shear thinning and shear thickening under shear

silica particles dispersed in ethylene glycol. The STF-treated Kevlar fabric provides almost identical protective behavior while being more flexible and thinner in comparison to untreated fabrics of the same weight. It is important to note that the improved ballistic performance is not solely attributed to the influence of the carrier liquid. This is evident from the experiments that Kevlar treated solely with ethylene glycol displays relatively inferior ballistic performance in comparison to both STF-treated and untreated fabrics. Additionally, it has been found that the yarn pull-out force of Kevlar fabrics treated with multi-phase STF comprising graphene oxide and carbon nanotubes was nearly twice that of pure Kevlar [20]. Furthermore, it has been observed that the incorporation of 34.89 wt.% STF (STKF) into Kevlar fabric results in a significant enhancement of stab resistance when compared to ten layers of pure Kevlar fabric [21]. This improvement is particularly noteworthy when eight layers of STKF are utilized. Additionally, SSG, a soft polymer derived from polyborosiloxane, exhibits a shear stiffening effect similar to STF [22]. With an increase in shear frequency, its storage modulus rises by 3–4 orders of magnitude [23]. Importantly, the solid SSG is particularly advantageous in protecting STF from heat and moisture. Combining Kevlar with SSG, the composite material demonstrates a 50% increase in penetration impact energy when compared to pure Kevlar [24].

3.5 STF-Based TENGs

Wearable TENGs in the market are ineffective under impact loadings, as they are unable to harvest high-speed kinetic energy or detect consecutive impact as self-powered sensors. Although some multifunctional TENGs possess excellent self-healing behavior, allowing them to recover the mechanical and triboelectric properties after destruction, they are still unable to withstand violent impacts and harvest energy. Therefore, it may be feasible to design all-in-one TENGs that offer higher protection and the ability to collect various types of impact energies by developing STF/SSG-based TENGs. In recent years, an increase in research has been observed. The growth in publications concerning the TENG and STF over the previous ten years is shown in Fig. 3.5.

3.5.1 Fabrication Methods and Working Mechanism

A large number of researches on textile-based TENGs are available, but a very limited number of researches have been conducted on the fabrication of STF-based TENGs with enhanced impact properties. Wang et al. [25] developed multifunctional TENG through the assembly of aluminum foil with a hybrid polymer matrix. Figure 3.6 shows the schematic representation of the fabrication procedure and working principle of the developed TENG. First, polytetrafluoroethylene (PTFE) nanoparticles and a small amount of polydimethylsiloxane (PDMS) were

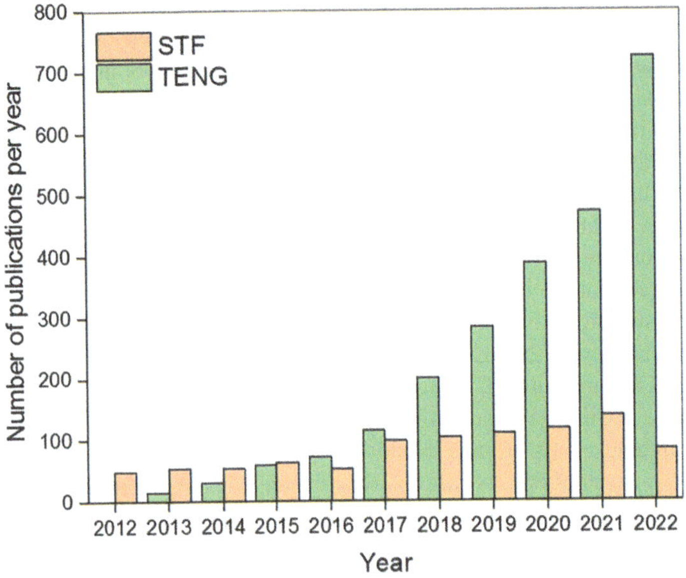

Fig. 3.5 The number of publications related to STFs and TENG based on the Scopus database

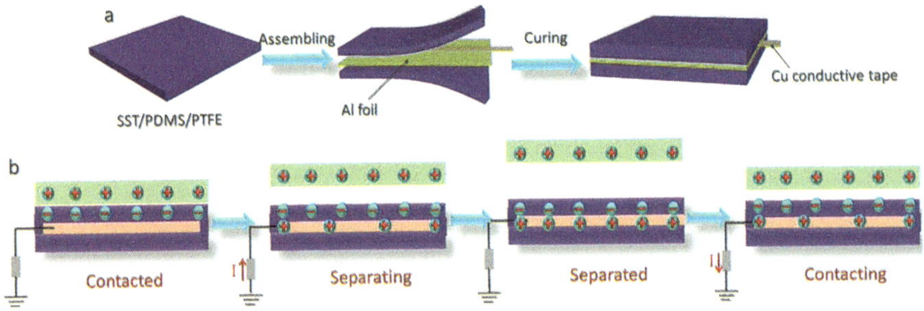

Fig. 3.6 (**a**) Schematic representation of fabrication procedure and (**b**) working principle of STF-based TENG [25]

incorporated into SSG. The TENG is then produced by sealing aluminum foil with a polymer matrix, consisting of SSG, PDMS, and PTFE. The initial contact induces charge transfer which is induced at the interface of human skin and TENG. This phenomenon is attributed to the different electron affinity of the two materials, which generates negative triboelectric charges on TENG and positive charges on human skin upon initial contact. Subsequently, when the positive skin finger leaves, a potential difference between the aluminum foil embedded in SSG/PDMS/PTFE and the ground is generated, which causes free electrons to flow. The system reaches electrostatic equilibrium upon leaving the human skin far away. The higher potential in aluminum foil can cause free electrons to flow back from the ground once the human finger approaches SST/PDMS/PTFE once more. This creates a negative current that continues until the system reverses to the initial state. The developed TENG can harvest kinetic energy sustainably with cyclic contacting-separating processes.

In another study, a conductive single electrode was created by combining the STF with an NaCl electrolyte [26]. In order to surround the aforementioned conductive STF in an exoskeleton, a silicon rubber with skin-friendliness and biocompatibility was used as the exterior material. The researchers directly attached the developed STF-based TENG to a desired location. Figure 3.7 shows the fabrication procedure and working principle of the STF-based TENG. First, the elastomer base and curing agent were combined in a 1:1 volume ratio to create the liquid silicon rubber. The resultant mixture was subsequently poured into a pre-fabricated mold to form a container. The silicon rubber was peeled off the mold after curing at room temperature for 3 h. Similar steps were taken to prepare an upper plate made of the silicon rubber after it had cured and served as a lid to contain the STF. The STF was then injected into the silicon rubber, and an aluminum wire was inserted into the STF. The container was then fitted with the lid as an upper cover to seal it. The seam line between the container and the lid was filled after another layer of liquid silicon rubber was applied over it. A porcine leather (2 × 2 cm) was used to simulate human skin as a counter electrode contacting the lid of the STF-TENG (4 × 4 cm).

The developed STF-based TENG operates on the basis of contact electrification and electrostatic induction. The working mechanism of the STF-based TENG is shown in four steps. (I): Through continuous contact, the porcine leather acquires a positive charge while the silicon rubber lid acquires a negative charge in accordance with the triboelectric series (Fig. 3.3). It is noteworthy that the inherent deformability of the lid enables an enlarged contact area. The liquid-like behavior of STF can be easily adapted to the shape of the porcine leather. (II): Positive charges are induced on the STF electrode by moving the porcine leather upward, resulting in the electron flow from the STF electrode to the ground. (III): Electrons stop flowing when the porcine leather moves far away and stops. The negative charges on the silicon rubber lid are fully balanced with the positive charges on the STF electrode. (IV): Electrons flow back from the STF electrode to the ground upon moving the porcine leather closer to the silicon rubber lid once again. Having a vertical movement in the porcine leather, the developed STF-based TENG generates an alternating current.

The STF-based TENG with high-performance energy harvesting and anti-impact effects was developed by Wang et al. [27]. The fabrication procedure involved the uniform mixing of ecoflex and carbonyl iron (CI), which were subsequently dropped into a mold. The film shell was removed and assembled with STF and PDMS/MWCNT/CI following a 20-min curing period at 50 °C. Structural depression was formed on ecoflex/CI by partially compressing air bubbles. As depicted in Fig. 3.8a, both air and STF were encapsulated in TENG. A good conductive path was formed by the evenly dispersed MWCNT, which was advantageous for conductivity. The working principle of the developed STF-based TENG is illustrated in Fig. 3.8b. The upper ecoflex/CI film was brought into contact with a conductive PDMS layer, resulting in the transfer of electrons between the surfaces (state I). Upon the

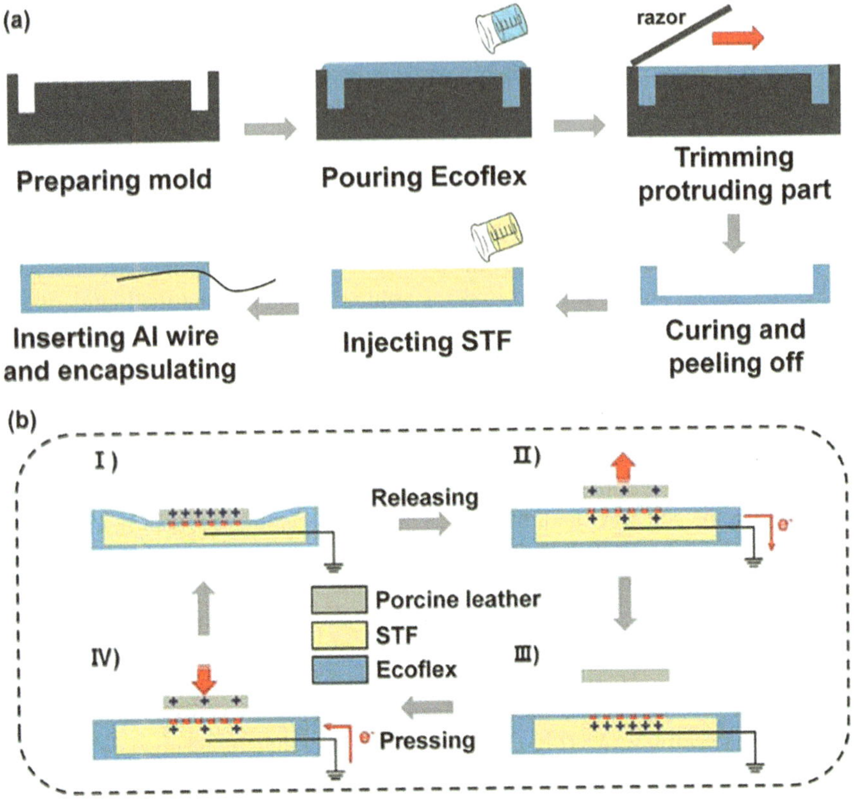

Fig. 3.7 (a) Schematic representation of the fabrication procedure and (b) working mechanism of the STF-based TENG [26]

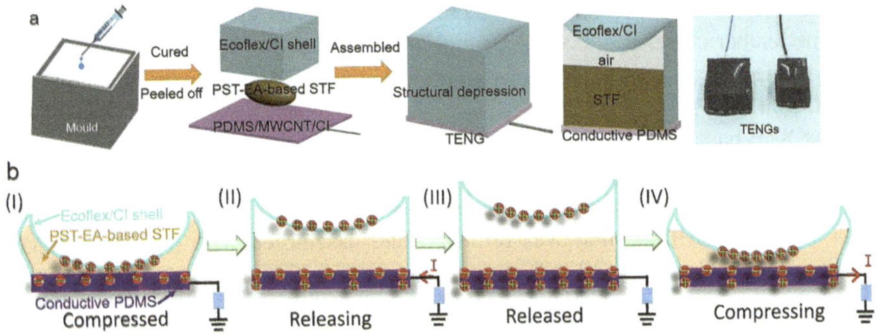

Fig. 3.8 (a) Schematic representation of fabrication procedure and (b) working principle of STF-based TENG [27]

departure of the positively charged ecoflex/CI, excess electrons on the PDMS layer created a potential difference between the polymer and the ground, leading to the flow of electrical current (state II). Once all the free electrons had dissipated, the system reached equilibrium and no further current could be detected (state III). Upon the return of the ecoflex/CI to the conductive PDMS layer, the high potential on the backside of the PDMS facilitated the flow of electrons back from the ground, resulting in the generation of an opposing current signal until the triboelectric nanogenerator (TENG) returned to its original state (state IV). This process enabled the ST-based TENG to effectively harvest and convert the mechanical energy into electrical energy.

Kim et al. [28] also designed a flexible and robust STF-based TENG to detect body motions including walking, running, and jumping. The developed system was made of an aluminum electrode, a rubber container, and cornstarch-based STF. To further enhance the triboelectricity of STF-based TENG, they incorporated polyvinylidene fluoride (PVDF) powders into the device. Figure 3.9 shows the schematic working mechanism of the developed STF-based TENG. The balanced electrical equilibrium is disrupted when the aluminum electrode is separated from the STF for the restoration of the rubber container because the impact of the negative charges on the surface of the STF starts to weaken. Consequently, electrons are transferred from the ground to the aluminum electrode to restore the electrical equilibrium between the surface of the Al_2O_3 layer and the aluminum electrode. The electrical equilibrium is restored once the restoration is complete. Upon application of external force, the effect of negative charges on the surface of the STF appears on the aluminum electrode, causing electrons to escape from the electrode to the ground to maintain the electrical equilibrium. As a result, the fabricated STF-based TENG serves as the source of AC electricity.

Textile/STF-based TENG was also developed by researchers by combining STF with Kevlar fabric to enhance the impact-resistant properties of wearable TENGs. Wang et al. [29] developed textile/STF-based TENG with excellent safeguarding and self-powered sensing by including graphene-based shear thickening materials (STF and SSG) into high-performance Kevlar fabric. Figure 3.10a shows the schematic demonstration of the fabrication procedure. Kevlar fabric was submerged in the SiO_2-based STF. The surface of the Kevlar/STF composite was then covered

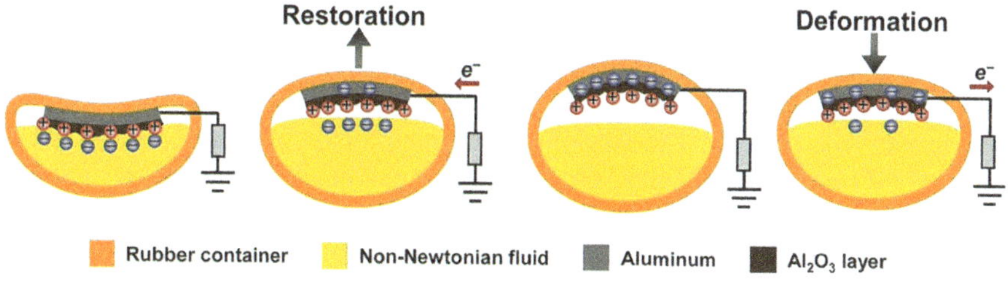

Fig. 3.9 The working mechanism of developed STF-based TENG [28]

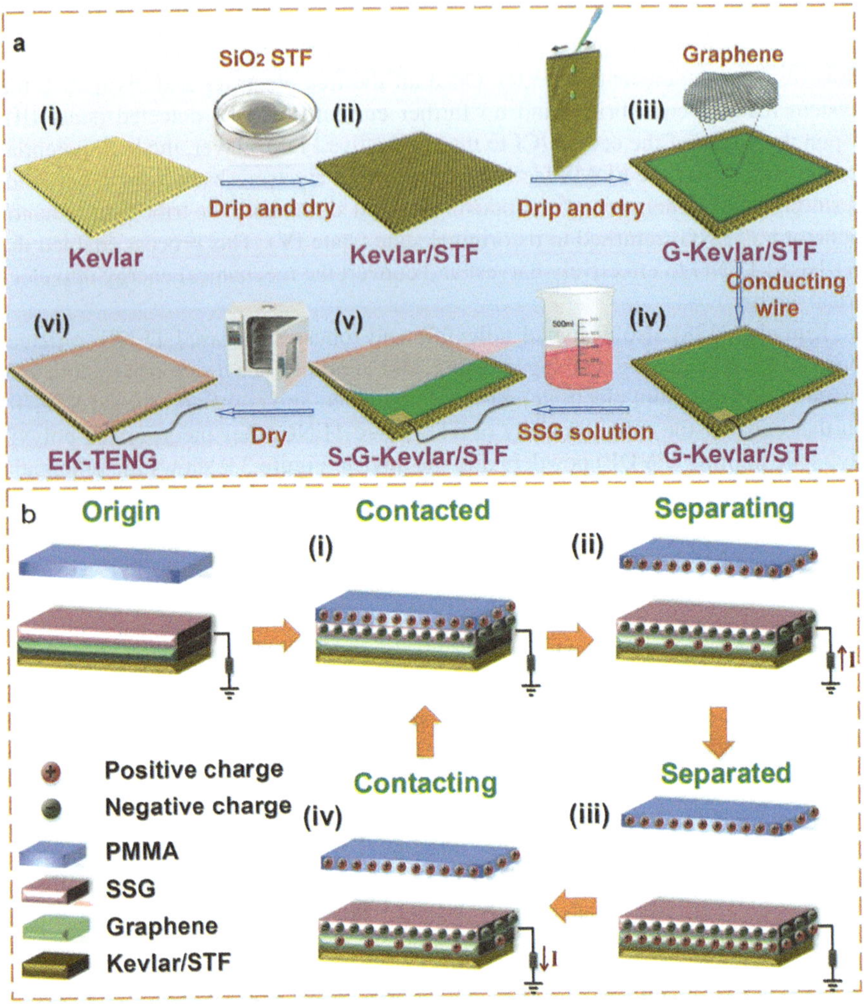

Fig. 3.10 (**a**) Schematic representation of fabrication procedure and (**b**) working mechanism of textile/STF-based TENG [29]

with a thin layer of the graphene-ethanol mixture. A conductive wire was then affixed onto the graphene layer, followed by the deposition of an SSG solution to form a thin film. Finally, the textile/STF-based TENG was obtained after drying the composites in a furnace. The working mechanism of the developed textile/STF-based TENG is shown in Fig. 3.10b. The coupling effect of triboelectrification and electrostatic induction was responsible for the functioning of the triboelectric transducing mechanism. The poly(methyl methacrylate) plate (PMMA) and textile/STF-based TENG were separated away in the original state. Upon contact, electrons were transferred from the interface of PMMA to textile/STF-based TENG due to their differing electron affinity, resulting in positive and negative triboelectric charges on the PMMA surface and textile/STF-based TENG surface, respectively (state I). As the PMMA left away, the potential difference between two surfaces caused the free electrons to flow from the electrode to the ground (state II). When the PMMA was separated from the system far away (state III), the system reached electrostatic equilibrium. Conversely, when the PMMA approached again, a current in opposite direction was obtained until the system returned to its initial state (state IV) [29].

Zhou et al. [30] also developed flexible TENG based on Kevlar fiber and conductive SSG. The conductive SSG was prepared by introducing a specific amount of CNT into the SSG. Then using a metal mold, conductive SSG was formed in a rectangular shape (Fig. 3.11a). A conductive wire was clamped between the two sticks of the conductive SSG. Finally, the wearable TENG was constructed by stacking two Kevlar fabrics with a layer of conductive SSG. Figure 3.11b illustrates how the developed Kevlar/SSG-based TENG harvests energy. Due to the different electron affinities, the interface of PDMS and TENG initially induced charge transferred according to the coupling effect of triboelectrification and electrostatic induction. It produced negative triboelectric charges on TENG while positive ones on PDMS (state I). As the positive PDMS moves away, a potential difference between TENG and the ground is created, resulting in the flow of free electrons (state II). When PDMS left far away and there were no freer electrons, the system reached electrostatic equilibrium (state III). The system turned into the initial state once PDMS approached Kevlar again, which caused a negative current signal. The high potential in SSG could also cause the free electrons to flow back from the ground (state IV).

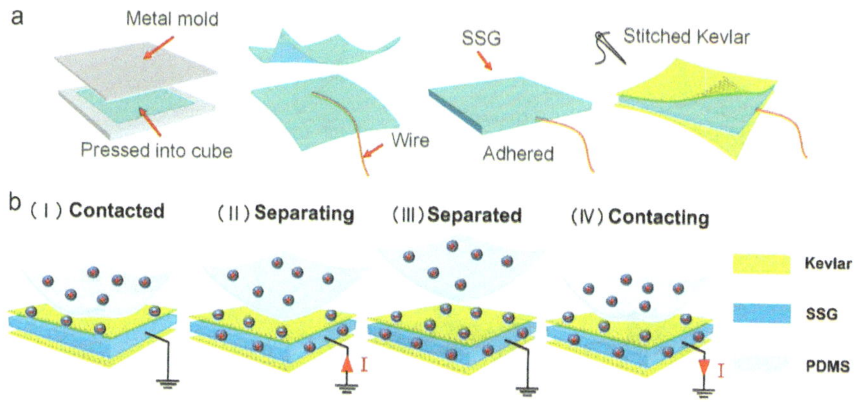

Fig. 3.11 (**a**) Schematic representation of fabrication procedure and (**b**) working mechanism of Kevlar/SSG-based TENG [30]

3.5.2 Triboelectric Performance of STF-Based TENGs

In comparison to conventional "solid-solid" TENGs, "liquid-solid" TENGs possess advantages such as longer lifespan due to the slight frictional interaction between the interfaces. As a result of the liquid state of STF, it also flew fluidly, causing electrons to transfer between its inner sidewalls. In the most relevant research in this field, Wang et al. [27] studied the resistance-dependent triboelectric behavior of a STF-based TENG in oscillator compression mode. The results indicate that the voltages exhibited a sharp increase when the external load resistance exceeded 105 Ω, reaching a maximum value of 23.59 V (Fig. 3.12a). Conversely, the current demonstrated an inverse relationship with resistance, decreasing as resistance increased. Figure 3.12b shows that the output power initially enhanced with resistance and then began to reduce after reaching a maximum level. They showed that the developed STF-based TENG enabled the harvesting of compression mechanical energy and functioning as a power source to illuminate LEDs. Figure 3.12c shows the experimental setup by loading on STF-based TENG with vertical swing excitation. The TENG was mounted on the end of the cantilever, and the stimuli were loaded using an oscillator. Enclosed STFs were set into motion, causing frictional interaction with the polymers and resulting in electron transfer and current output (as shown in Figs. 3.12d and e). The voltage output was observed to increase from 11.3 mV at a swaying amplitude of 1 cm to 46.9 mV at an amplitude of 8 cm (Fig. 3.12f). Therefore, the developed STF-based TENG demonstrated the ability to harvest and convert mechanical energy from STF-flow into electricity when operating in non-contacted mode.

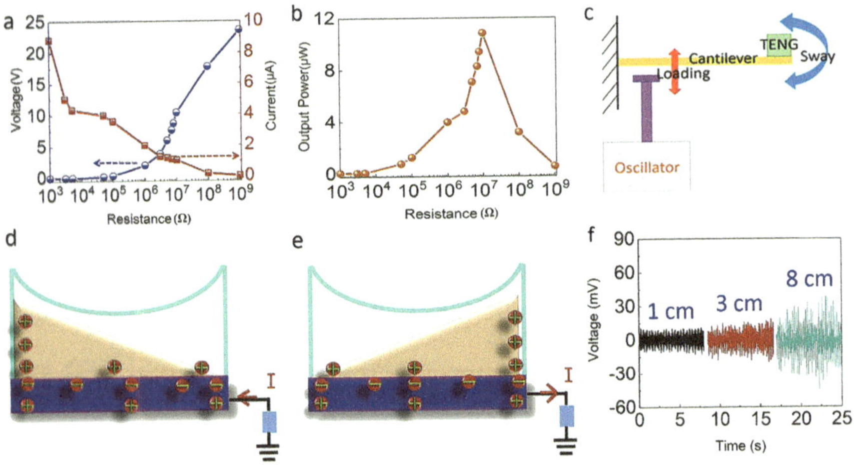

Fig. 3.12 (**a**) Triboelectric voltage, current, and (**b**) output power, (**c**) the experimental setup for vertical swing excitation, (**d**, **e**) electrons transferred during the STF flowing process, and (**f**) the corresponding voltages of STF-based TENG [27]

The triboelectric performance of textile/STF-based TENG were analyzed by using an oscillator system [29]. The output voltages of textile/STF-based TENG under various applied forces at a frequency of 10 Hz (Fig. 3.13a) demonstrated an increasing trend as the applied forces increased. The voltage signals of textile/STF-based TENG at varying input frequencies (Fig. 3.13b) indicate a consistent increasing trend, which can be attributed to the shorter separation time and the accumulation of more generated charges on the electrode. Furthermore, the current, maximum voltage, and power density of textile/STF-based TENG were determined by changing the external resistance at impact force of 40 N and frequency of 10 Hz as shown in Fig. 3.13c. The highest peak power density of 25.8 mW/m^2 was achieved at 10 MΩ. The electrical stability of textile/STF-based TENG was evaluated by conducting 1000 cycles of loading-unloading stimulations (Fig. 3.13d and f) and the results indicate excellent stability.

Zhou et al. [30] also investigated the resistance-dependent triboelectric properties of developed Kevlar/SSG-based TENG. As depicted in Fig. 3.14a, the Kevlar/SSG-based TENG was affixed to a flat surface and subjected to excitations by an oscillator. The loading force was measured by a force sensor that was placed close to the oscillator's probe. The voltage signal measured by the multimeter increased from 6 mV to 24 V as the external resistance increased from 1 kΩ to 1 GΩ and the current reduced from 6 to 0.02 μA (Fig. 3.14b). The highest peak power density of 5.3 mW/m^2 was achieved with a voltage of 13.1 V under 9 MΩ (Fig. 3.14c). The Kevlar/SSG-based TENG could serve as a novel power source to illuminate LED arrays, as shown in Fig. 3.14d. Furthermore, the output properties of Kevlar/SSG-based TENG were positively influenced by the constant frequency of contact-separation loading.

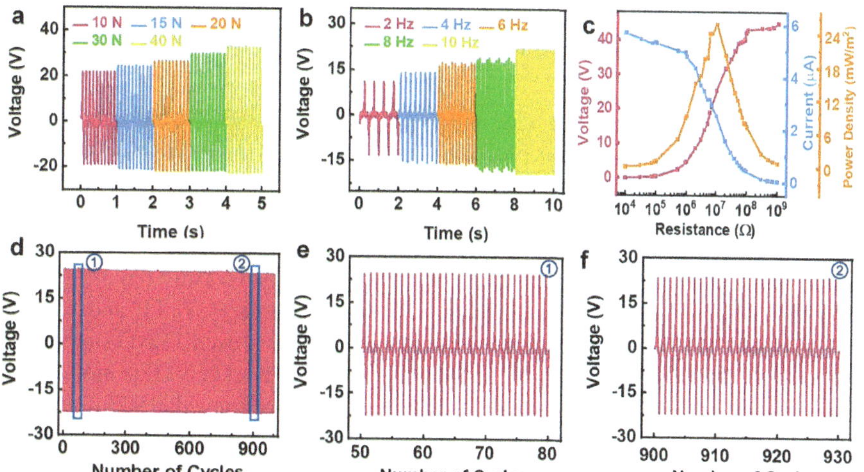

Fig. 3.13 (**a**) Force-dependent output voltages at 10 Hz, (**b**) frequency-dependent output voltages at 10 N, (**c**) voltage, current, and output power under various external resistances, (**d–f**) cycling stability of textile/STF-based TENG at 10 MΩ [29]

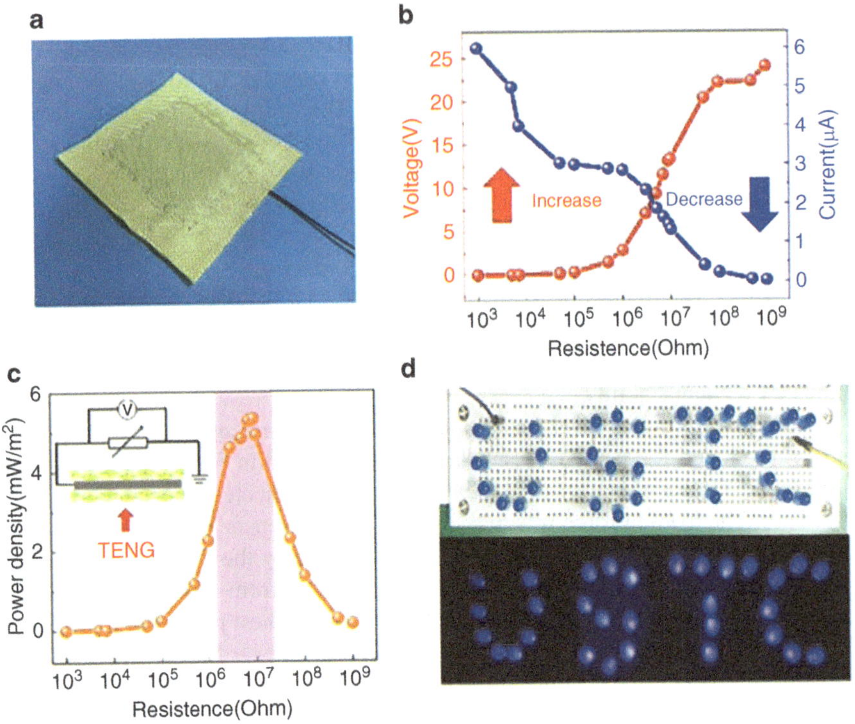

Fig. 3.14 (**a**) The as-designed Kevlar/SSG-based TENG, (**b**) voltage, current, and (**c**) output power under various external resistances, and (**d**) Kevlar/SSG-based TENG harvested mechanical energy and lit up LEDs [30]

3.5.3 Impact-Absorbing Performance of STF-Based TENGs

The impact-absorbing performance and mechanical robustness with the STF-based TENGs are important for their practical application. Previous studies have shown the significant role of STF in enhancing the impact energy absorption of materials [31–34]. It is worth mentioning that the impact-absorbing performance of the STF-based TENGs depends on the thickness of STF. To study the effect of STF thickness on the impact-absorbing performance of STF-based TENGs, Yun et al. [26] fabricated STF-based TENGs with different STF thicknesses of 3, 7, 11, and 15 mm. They showed that as the STF thickness increased, the STF-based TENG suppressed more force from the impact. Their results suggest that a thicker STF is more desirable for effective impact-absorbing. Nevertheless, augmenting the STF thickness results in a corresponding increase in the bulkiness and weight of the STF-based TENG, rendering it unsuitable for wearable applications. It has been determined that an STF thickness of 7 mm is the optimal value for achieving both mobility and impact-absorption capabilities, thereby ensuring comfortable wearability. The cyclic durability test also validated the capability of STF-based TENG to produce electrical energy with long-term durability and structural reliability, even under conditions of heavy impact.

Kevlar, as one of the most well-known high-performance textile materials, has been utilized in body armor due to its high modulus and exceptional puncture resistance. Kevlar has been shown to have better anti-impact performance when combined with shear thickening materials. Hence, several studies have been carried out to further explore the protective as well as impact energy-harvesting capabilities of textile/STF-based TENGs. For instance, the energy-absorbing property of textile/STF-based TENG under low-velocity impact was evaluated using the drop weight test [29]. It is found that under the same experimental conditions, the peak impact force exerted on a 30-layer textile/STF-based TENG composite was observed to decrease to 439 N, representing a reduction of 52.3% in comparison to neat Kevlar. The textile/STF-based TENG also demonstrated higher force dissipation under impacts. Moreover, textile/STF-based TENG demonstrated the ability to generate voltage signals through the self-powered sensing effect. The impact resistance of textile/STF-based TENG was also studied by firing a spherical projectile from a gas gun [29]. The results of ballistic impact experiments showed that the toy wearing Kevlar-suit suffered significant injury as it was fully penetrated, whereas the textile/STF-based TENG suit effectively blocked the projectile. A longer projectile breakdown time and a larger deformation of textile/STF-based TENG were obtained than Kevlar. This experiment serves as evidence that the textile/STF-based TENG can impede damage caused by high-speed shooting.

Zhou et al. [30] studied the anti-impact mechanical performance of flexible TENG based on Kevlar fiber and conductive SSG. They fixed the TENG on the clamp and connected a conductive wire to another oscilloscope to collect the electrical signal generated by the TENG during impact. The SSG-TENG demonstrated resistance to ballistic shooting at a velocity of 76.3 m/s, resulting in a single voltage peak of 0.6 V due to the contact between the projectile and the Kevlar/SSG-based TENG. However, when the projectile had a velocity of 114 m/s, the Kevlar/SSG-based TENG was completely penetrated, resulting in a significantly different voltage-time curve. As the projectile made contact with the Kevlar/SSG-based TENG and started to pierce the fiber, a positive voltage of 2 V was produced. It is evident that the impact velocity had a positive effect on the output voltages, and the electrical curves could be utilized to evaluate external impact and determine the damage sustained by the Kevlar/SSG-based TENG. It is noteworthy that the developed Kevlar/SSG-based TENG could continue to gather energy and respond to shocks even after being penetrated.

3.6 Practical Applications of STF-Based TENGs

Many researchers have been interested in the practical applications of TENGs, due to their advantages of simplicity, cost-effectiveness, and industrial availability. The potential applications of TENGs in energy and sensors have already been done. This section aims to provide a brief overview of the applications of flexible STF-based TENGs, highlighting a few examples of impact energy harvesting systems.

Figure 3.15a shows a human hand array assembled with STF-based TENGs [27]. It can be worn on the hand comfortably, and thanks to its high flexibility, it can adapt to intricate mechanical deformations. Figure 3.15b and c show the sensing capability of STF-based TENG to the bending excitation. The array of STF-based TENG on the back of the hand also allowed for the perception and display of different compression forces (Fig. 3.15d and e). The hand array, possessing heightened sensitivity and self-sustaining real-time monitoring capabilities, has the potential to be utilized in the development of intelligent robots, wearable healthcare devices, and safety alert systems. In order to simulate the impact-absorbing effect, a stone was used to compress the Kevlar/STF-based TENG pad from various falling heights, while a force sensor was employed to record the resulting impact force. Notably, the forces dissipated by the Kevlar/STF-based TENG pad were significantly lower than those directly loaded onto the force sensor, indicating that the designed array is

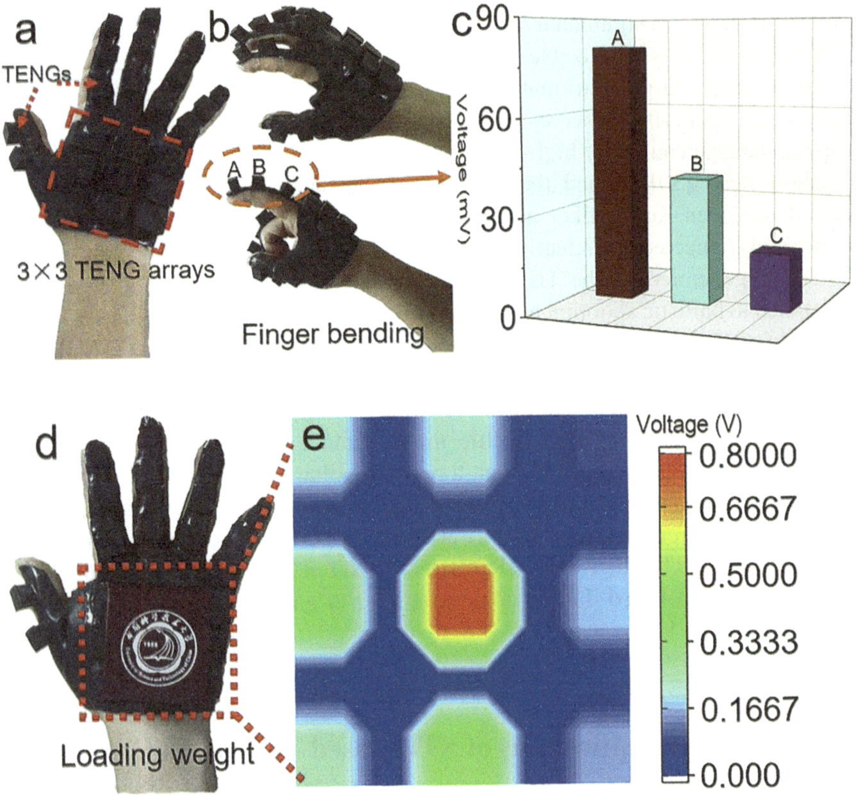

Fig. 3.15 (**a**) A human hand array assembled with STF-based TENG, (**b**) index finger bending, (**c**) the corresponding maximum voltages, (**d**, **e**) pressure distribution of the array on the back of hand [27]

highly efficient in impact energy absorption. Furthermore, the output voltages of the Kevlar/STF-based TENG pad were observed to increase from 0.84 to 2.18 V as impact height was changed from 10 to 30 cm, demonstrating a force-sensing effect. During seven cycles of loading-unloading impacts, the device also displayed highly reliable and stable mechanical characteristics. Kevlar/TENG as a functional and wearable electronics was able to more reliably protect users from emergent kinetic harm thanks to this advantageous anti-impact/multi-sensing property. Additionally, the performance of the self-powered monitoring system made it possible to detect and evaluate various tactile violations.

Figure 3.16 shows the potential of STF-based TENGs as an impact monitoring device, which was affixed to the knee of an individual [26]. The voltage signal that the STF-based TENG produced in response to an external impact was transmitted to an impact monitoring unit, connected to a smartphone. Figure 3.16b displays captured images of the monitoring system under various conditions, including no touch, weak press, strong impact, and weak press again. The output voltage of the STF-based TENG was 0.3 V and 13.0 V when a weak press and a strong impact were applied, respectively. A red sign warned about the danger when the output voltage by the external impact exceeded 2.0 V. The STF-based TENG exhibited high mechanical robustness and continued to function even after strong impacts. Furthermore, a message was delivered to a mobile device (Fig. 3.16c). The developed STF-based TENG demonstrated a great potential as an emergency alarm. For example, if a patient with this device is in danger while alone in a remote area, the device can identify the threat and send an alert to a guardian or a medical team requesting rescue, giving them enough time to perform a lifesaving action.

Wang et al. [29] further studied the potential of textile/STF-based TENG to transmit sensing signals to electrical devices and alarm human beings. A wireless smart passive alarm utilizing textile/STF-based TENG, which incorporates body motion monitoring and impact sensing capabilities for remote emergency rescue, has been successfully designed. The system was subjected to three distinct loading

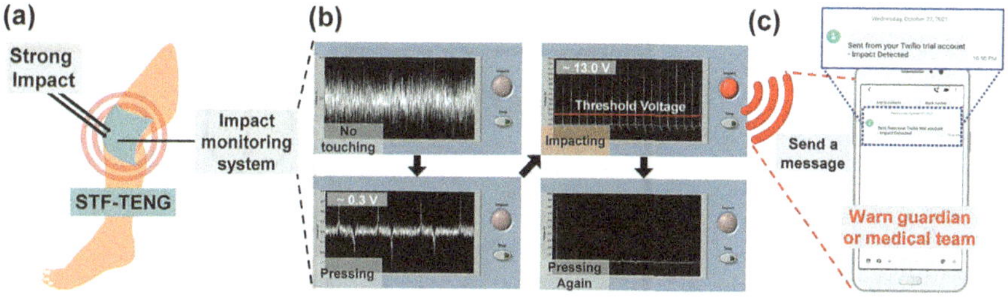

Fig. 3.16 (**a**) Schematic of the STF-based TENG attached to the knee with an applied external impact. The voltage signal generated by the STF-TENG is delivered to the impact monitoring system. (**b**) Images were recorded from the impact monitoring system in the following sequence: no touching, pressing, impacting, and pressing again. (**c**) A smartphone-displayed screenshot of a received message [26]

modes with various frequencies, including human walking, medium and high-velocity impacts to generate voltage signals.

Figure 3.17a shows the alarming system based on textile/STF-based TENG that was developed by Wang et al. [29]. The signals from textile/STF-based TENG were initially subjected to filtering and amplification by the signals processing circuits. After that, the signals were transformed by a converter from analog to digital form, and finally analyzed by a microcontroller. This unit was equipped with a visual screen, which enabled the dynamic display of the detected waveforms and the convenient setting of the threshold of the internal program and the alarm switch through a user friendly computer. A smartphone application was programmed to wirelessly transmit electrical signals to a smartphone via Bluetooth communication when the microcontroller unit detected a danger. The application program interface would prompt the "Alarm" through a dialog box (Fig. 3.17c).

The Kevlar/SSG-based TENG developed by Zhou et al. [30] presented an excellent demonstration as a self-powered sensor to monitor different body motions because it was soft and highly flexible (Fig. 3.18a). Figure 3.18b shows the typical voltage-time curves generated by the SSG-TENG during walking, jumping, and running. The peak voltages during walking and running were 0.5 and 0.7 V,

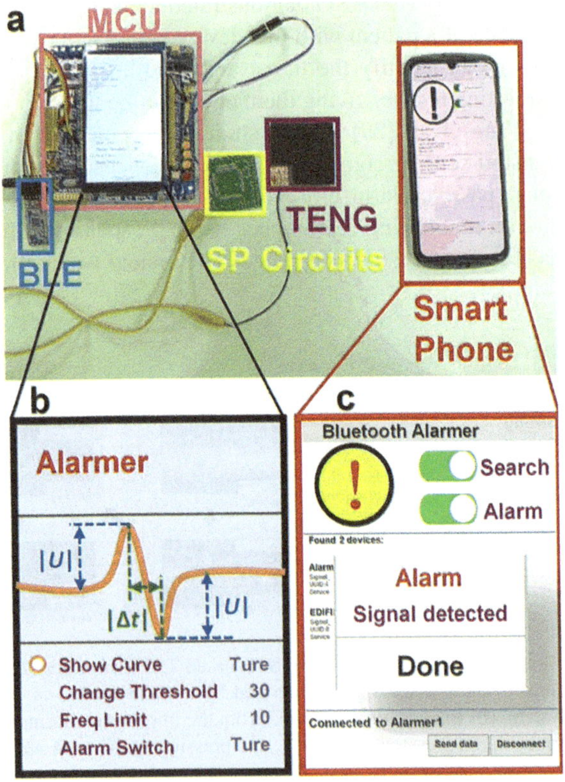

Fig. 3.17 (**a**) The sensing system contained TENG-based Bluetooth and the interface of (**b**) microcontroller and (**c**) smartphone [29]

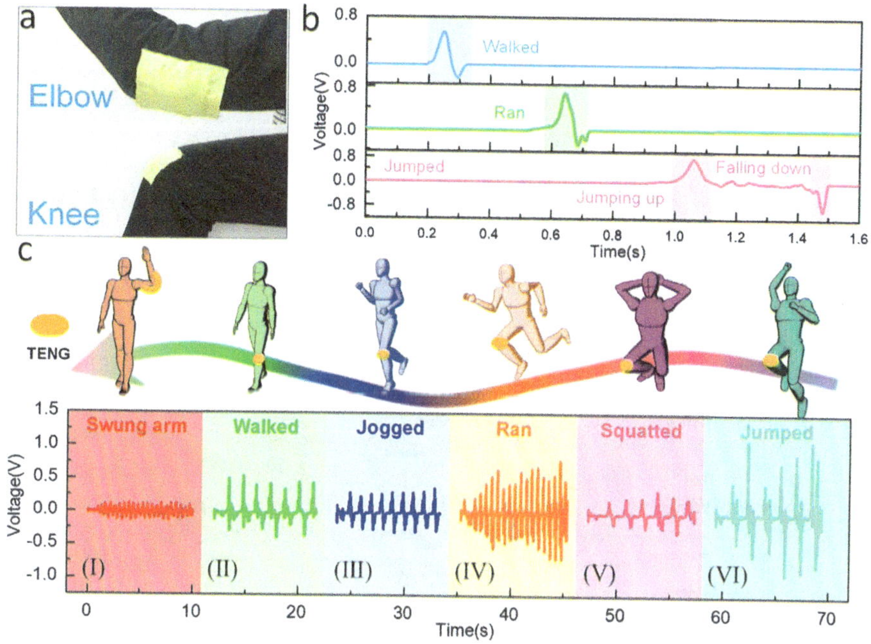

Fig. 3.18 (**a**) Kevlar/SSG-based TENG device could be worn on elbow and knee, (**b**) typical single voltage curves of walking, running, and jumping, and (**c**) detecting various human motions including (**c**) swing arm, walking, jogging, running, squatting, and jumping [30]

respectively, which were caused by the contact between Kevlar/SSG-based TENG and knee. When the person jumped, a positive voltage signal was generated, and when they hit the ground, a negative voltage peak was obtained. Notably, the small time period between the two voltage peaks can be utilized to evaluate the speed of the jump as well as the duration of the jump and the time spent falling. Furthermore, the Kevlar/SSG-based TENG demonstrated its reliability in detecting different joint motions (Fig. 3.18c). High sensitivity to various external excitations was another characteristic of developed Kevlar/SSG-based TENG. Figure 3.18c shows how it could capture the low frequencies of walking, jumping, and squatting. Moreover, the Kevlar/SSG-based TENG was capable of detecting higher-frequency movements, including jogging, running, and arm swinging.

Zhou et al. [30] also studied the potential application of developed Kevlar/SSG-based TENG in exercise monitoring. To detect human movements, the Kevlar/SSG-based TENG array made up of different units was attached to a shoe. First, walking was studied to gauge how well the Kevlar/SSG-based TENG array sensed it. Figure 3.19b shows that unit 7 of the array began to output a voltage of −11.1 V at 0.3 s as soon as the heel touched the ground. Then, as feet stepped on the ground, units 5, 4, and 2 gradually produced −7.0, −12.7, and − 6.3 V. The appearance of a positive signal on each unit demonstrated that the foot had actually lifted off the ground. This outcome demonstrated that the developed Kevlar/SSG-based TENG

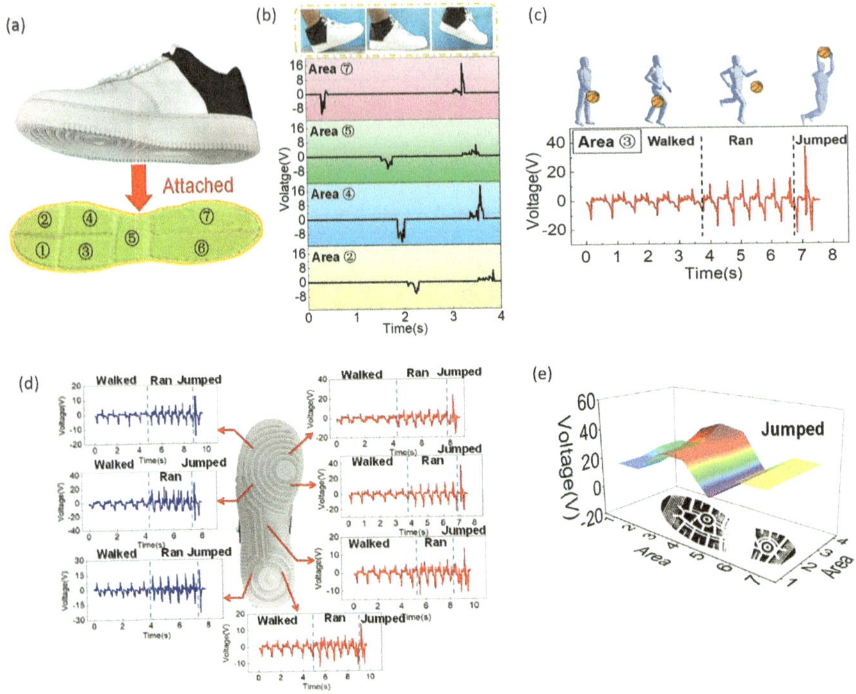

Fig. 3.19 (**a**) Kevlar/SSG-based TENG sole array on the shoe, (**b**) electrical signals outputted by different units of the array during one process of human walk, (**c**) typical voltage-time curves generated by the "area 3" of the array to monitor various foot movements, (**d**) all the units of the array could assess the walking, running, and jumping processes, and (**e**) voltage comparison of all units under jumping [30]

array was capable of accurately and consecutively detecting foot motions. Furthermore, unit 3 was able to sense the process of a basketball player shooting, which involved walking, jumping, and running (Fig. 3.19c). Running obviously has a higher voltage frequency than walking, and jumping has the highest voltage at its maximum. Other array units were also capable of detecting these series of movements in a similar manner (Fig. 3.19d). However, due to the different sole pressures, all units under the same stimulation had different output voltages. Figure 3.19e shows the diagram of the voltage distribution of the array during jumping. Clearly, unit 3 produced the highest voltage of 39.7 V due to the highest sustaining pressure during jumping.

3.7 Summary

In this chapter, the recent progress of shear thickening material-based TENGs with impact energy harvesting and safeguarding effects has been reviewed. The STF-TENG exhibited characteristics of a liquid at low-velocity contacts and solid-like behavior at high velocities. Furthermore, it demonstrated a notable escalation in electrical output as contact velocity increased, which was attributed to the increased viscosity resulting from the particle network formations in the STF. The working principle and fabrication methods of STF-based TENG have been introduced. The role of STF on energy-absorbing performance and output voltage performance of STF-based TENGs has been critically discussed. Finally, the potential of STF-based TENG in human healthcare application, remote emergency rescue, and impact monitoring system has been discussed. The impact monitoring system based on STF-TENG can detect a strong impact and issued a warning by means of an alert sign that blinked and an emergency message that was dispatched to a smartphone. Based on the considerable progresses that have already been achieved, the noteworthy breakthroughs in the near future are anticipated through the combination of highly efficient and flexible TENGs with shear thickening materials.

Acknowledgments This work was funded by the Scientific and Technological Research Council of Turkey (TUBITAK, Grant No. 1059B212200845).

References

1. Moghadam, B.H., Hasanzadeh, M., Simchi, A.: Self-powered wearable piezoelectric sensors based on polymer nanofiber-metal-organic framework nanoparticle composites for arterial pulse monitoring. ACS Appl. Nano Mater. **3**, 8742–8752 (2020). https://doi.org/10.1021/acsanm.0c01551
2. Fan, F.R., Tang, W., Wang, Z.L.: Flexible nanogenerators for energy harvesting and self-powered electronics. Adv. Mater. **28**, 4283–4305 (2016). https://doi.org/10.1002/adma.201504299
3. Wu, C., Wang, A.C., Ding, W., Guo, H., Wang, Z.L.: Triboelectric nanogenerator: a foundation of the energy for the new era. Adv. Energy Mater. **9**, 1–25 (2019). https://doi.org/10.1002/aenm.201802906
4. Dassanayaka, D.G., Alves, T.M., Wanasekara, N.D., Dharmasena, I.G., Ventura, J.: Recent progresses in wearable triboelectric nanogenerators. Adv. Funct. Mater. **32** (2022). https://doi.org/10.1002/adfm.202205438
5. Nurmakanov, Y., Kalimuldina, G., Nauryzbayev, G., Adair, D., Bakenov, Z.: Structural and chemical modifications towards high-performance of triboelectric nanogenerators. Nanoscale Res. Lett. **16**, 122 (2021). https://doi.org/10.1186/S11671-021-03578-Z
6. Hinchet, R., Seung, W., Kim, S.W.: Recent progress on flexible triboelectric nanogenerators for selfpowered electronics. ChemSusChem. **8**, 2327–2344 (2015). https://doi.org/10.1002/CSSC.201403481
7. Zhu, G., Pan, C., Guo, W., Chen, C.Y., Zhou, Y., Yu, R., Wang, Z.L.: Triboelectric-generator-driven pulse electrodeposition for micropatterning. Nano Lett. **12**, 4960–4965 (2012). https://doi.org/10.1021/NL302560K/SUPPL_FILE/NL302560K_SI_001.PDF

8. Yoon, H.J., Ryu, H., Kim, S.W.: Sustainable powering triboelectric nanogenerators: approaches and the path towards efficient use. Nano Energy. **51**, 270–285 (2018). https://doi.org/10.1016/J.NANOEN.2018.06.075
9. Zhong, J., Zhang, Y., Zhong, Q., Hu, Q., Hu, B., Wang, Z.L., Zhou, J.: Fiber-based generator for wearable electronics and mobile medication. ACS Nano. **8**, 6273–6280 (2014). https://doi.org/10.1021/nn501732z
10. Paosangthong, W., Torah, R., Beeby, S.: Recent progress on textile-based triboelectric nanogenerators. Nano Energy. **55**, 401–423 (2019). https://doi.org/10.1016/j.nanoen.2018.10.036
11. Hasanzadeh, M., Mottaghitalab, V., Sharma, P.: Puncture resistance enhancement of woven fabrics using concentrated nanosilica suspension. Procedia Eng. **173**, 1494–1498 (2017). https://doi.org/10.1016/j.proeng.2016.12.228
12. Laha, A., Majumdar, A.: Interactive effects of p-aramid fabric structure and shear thickening fluid on impact resistance performance of soft armor materials. Mater. Des. **89**, 286–293 (2016). https://doi.org/10.1016/j.matdes.2015.09.077
13. Sheikhi, M.R., Hasanzadeh, M.: Multi-phase shear thickening fluid, shear thick. Fluid., 33–51 (2023). https://doi.org/10.1007/978-3-031-25717-9_3
14. Sheikhi, M.R., Hasanzadeh, M., Gürgen, S.: Conductive shear thickening fluids for multifunctional purposes, shear thick. Fluid, 13–25 (2023). https://doi.org/10.1007/978-3-031-35521-9_2
15. Gürgen, S., de Sousa, R.J.A.: Rheological and deformation behavior of natural smart suspensions exhibiting shear thickening properties. Arch. Civ. Mech. Eng. **20** (2020). https://doi.org/10.1007/s43452-020-00111-4
16. Gürgen, S., Kuşhan, M.C., Li, W.: The effect of carbide particle additives on rheology of shear thickening fluids. Korea Aust. Rheol. J. **28**, 121–128 (2016). https://doi.org/10.1007/s13367-016-0011-x
17. Gürgen, S., Sofuoğlu, M.A., Kuşhan, M.C.: Rheological compatibility of multi-phase shear thickening fluid with a phenomenological model. Smart Mater. Struct. **28**, 035027 (2019). https://doi.org/10.1088/1361-665X/ab018c
18. Sheikhi, M.R., Hasanzadeh, M., Gürgen, S.: The role of conductive fillers on the rheological behavior and electrical conductivity of multi-functional shear thickening fluids (M-STFs). Adv. Powder Technol. **34**, 104086 (2023). https://doi.org/10.1016/j.apt.2023.104086
19. Lee, Y.S., Wetzel, E.D., Wagner, N.J.: The ballistic impact characteristics of Kevlar® woven fabrics impregnated with a colloidal shear thickening fluid. J. Mater. Sci. **38**(13), 2825–2833 (2003). https://doi.org/10.1023/A:1024424200221
20. Liu, L., Cai, M., Liu, X., Zhao, Z., Chen, W.: Ballistic impact performance of multi-phase STF-impregnated Kevlar fabrics in aero-engine containment. Thin-Walled Struct. **157**, 107103 (2020). https://doi.org/10.1016/J.TWS.2020.107103
21. Qin, J., Guo, B., Zhang, L., Wang, T., Zhang, G., Shi, X.: Soft armor materials constructed with Kevlar fabric and a novel shear thickening fluid. Compos. Part B Eng. **183**, 107686 (2020). https://doi.org/10.1016/j.compositesb.2019.107686
22. Gürgen, S., Yıldız, T.: Stab resistance of smart polymer coated textiles reinforced with particle additives. Compos. Struct. **235**, 111812 (2020). https://doi.org/10.1016/J.COMPSTRUCT.2019.111812
23. Zhang, S., Wang, S., Hu, T., Xuan, S., Jiang, H., Gong, X.: Study the safeguarding performance of shear thickening gel by the mechanoluminescence method. Compos. Part B Eng. **180**, 107564 (2020). https://doi.org/10.1016/j.compositesb.2019.107564
24. Wang, S., Xuan, S., Liu, M., Bai, L., Zhang, S., Sang, M., Jiang, W., Gong, X.: Smart wearable Kevlar-based safeguarding electronic textile with excellent sensing performance. Soft Matter. **13**, 2483–2491 (2017). https://doi.org/10.1039/C7SM00095B
25. Wang, S., Ding, L., Wang, Y., Gong, X.: Multifunctional triboelectric nanogenerator towards impact energy harvesting and safeguards. Nano Energy. **59**, 434–442 (2019). https://doi.org/10.1016/j.nanoen.2019.02.060

26. Yun, S.Y., Tcho, I.W., Kim, W.G., Kim, D.W., Son, J.H., Lee, S.W., Choi, Y.K.: Mechanically robust triboelectric nanogenerator with a shear thickening fluid for impact monitoring. J. Mater. Chem. A. **10**, 10383–10390 (2022). https://doi.org/10.1039/d2ta01209j
27. Wang, S., Liu, S., Zhou, J., Li, F., Li, J., Cao, X., Li, Z., Zhang, J., Li, B., Wang, Y., Gong, X.: Advanced triboelectric nanogenerator with multi-mode energy harvesting and anti-impact properties for smart glove and wearable e-textile. Nano Energy. **78**, 105291 (2020). https://doi.org/10.1016/j.nanoen.2020.105291
28. Kim, Y., Yun, J., Kim, D.: Robust and flexible triboelectric nanogenerator using non-Newtonian fluid characteristics towards smart traffic and human-motion detecting system. Nano Energy. **98**, 107246 (2022). https://doi.org/10.1016/j.nanoen.2022.107246
29. Wang, W., Zhou, J., Wang, S., Yuan, F., Liu, S., Zhang, J., Gong, X.: Enhanced Kevlar-based triboelectric nanogenerator with anti-impact and sensing performance towards wireless alarm system. Nano Energy. **91**, 106657 (2022). https://doi.org/10.1016/j.nanoen.2021.106657
30. Zhou, J., Wang, S., Yuan, F., Zhang, J., Liu, S., Zhao, C., Wang, Y., Gong, X.: Functional Kevlar-based triboelectric nanogenerator with impact energy-harvesting property for power source and personal safeguard. ACS Appl. Mater. Interfaces. **13**, 6575–6584 (2021). https://doi.org/10.1021/acsami.0c18308
31. Gürgen, S., Kuşhan, M.C., Li, W.: Shear thickening fluids in protective applications: a review. Prog. Polym. Sci. **75**, 48–72 (2017). https://doi.org/10.1016/j.progpolymsci.2017.07.003
32. Hasanzadeh, M., Mottaghitalab, V., Rezaei, M., Babaei, H.: Numerical and experimental investigations into the response of STF-treated fabric composites undergoing ballistic impact. Thin-Walled Struct. **119**, 700–706 (2017). https://doi.org/10.1016/j.tws.2017.07.020
33. Hasanzadeh, M., Mottaghitalab, V.: The role of shear-thickening fluids (STFs) in ballistic and stab-resistance improvement of flexible armor. J. Mater. Eng. Perform. **23**, 1182–1196 (2014). https://doi.org/10.1007/s11665-014-0870-6
34. Hasanzadeh, M., Mottaghitalab, V., Babaei, H., Rezaei, M.: The influence of carbon nanotubes on quasi-static puncture resistance and yarn pull-out behavior of shear-thickening fluids (STFs) impregnated woven fabrics. Compos. Part A Appl. Sci. Manuf. **88**, 263–271 (2016). https://doi.org/10.1016/j.compositesa.2016.06.006

Chapter 4
Shear Thickening Electrolytes for Lithium-Ion Batteries

Mahdi Hasanzadeh

4.1 Introduction

The rechargeable batteries have garnered significant interest from both researchers and industries. Lithium-ion batteries (LIBs) maintain a dominant position in the market. LIBs have been utilized in large-scale applications, such as electric vehicles, aerospace, cell phones, and computers. Nevertheless, the widespread dissemination and utilization of electric vehicles have encouraged researchers to focus on collision safety performance, specifically in terms of mechanical abuse. Commercial LIBs typically consist of a porous separator, positive and negative electrodes, and an electrolyte. Recently, incidents of battery-related explosions and fires have still been observed in electric vehicle accidents, leading to significant public concerns. These phenomena are often attributed to internal thermal runaway resulting from mechanical damage. When a battery is exposed to external loading, the interaction between oxygen, fuel, and heat can cause fire or explosion, threatening life safety. Figure 4.1 shows the three stages of the thermal runaway process of batteries. Upon exposure to an external high-velocity impact, experiences an elevation in temperature and subsequent thermal contraction of the separator, thereby initiating an internal short circuit (stage I). Subsequently, the battery temperature rises unusually, accompanied by an exothermic reaction (stage II). Consequently, the flammable and volatile electrolyte comes in contact with air, leading to a thermal runaway at high temperatures. This thermal runaway can then trigger a spark, a fire, or even an explosion within the system (stage III) [1].

In order to address the issue of fire and explosion, resulting from mechanical abuse in batteries, various approaches have been commonly employed to enhance their safety performance. These approaches include the use of flame-retardant

M. Hasanzadeh (✉)
Department of Textile Engineering, Yazd University, Yazd, Iran
e-mail: m.hasanzadeh@yazd.ac.ir

S. Gürgen (ed.), *Smart Systems with Shear Thickening Fluid*,
https://doi.org/10.1007/978-3-031-53570-3_4

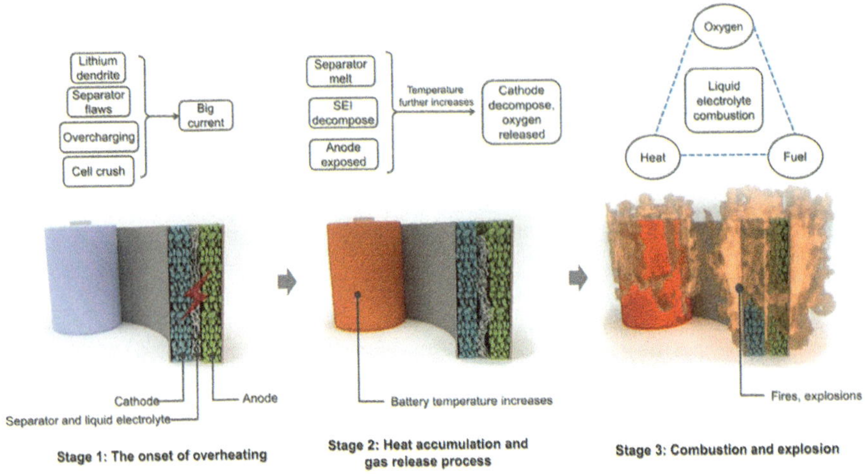

Fig. 4.1 Stages in thermal runaway of batteries [1]

materials, protective packages, and stable electrolytes, particularly solid-state electrolytes and ionic liquids. The incorporation of flame retardant additives, such as organic phosphate compounds, confers advantageous flame-resistant characteristics upon electrolytes. Nevertheless, this inclusion also causes an elevation in electrolyte viscosity and a concomitant reduction in electrochemical performance. Physical protection packaging is an alternative approach, which results in increased weight and additional strain on the battery. Moreover, it results in a decrease in the energy density of the battery and an increase in the cost of packaging. The utilization of ionic liquids is advantageous due to their nonflammable nature. However, there are certain drawbacks associated with their use, such as high costs, low lithium transfer numbers, and low multiplicative performance, which necessitate resolution. Conventional protection methods typically compromise partial electrochemical performance and result in unnecessary waste. Consequently, it is preferable to attain impact resistance by substituting current battery components with alternatives that possess improved mechanical strength [1].

The exploration of novel electrolytes, which possess anti-impact properties and maintain cell function, for the construction of safer and more sustainable LIBs has emerged as a key challenge for future advancements. Numerous solid nonflammable electrolytes with high mechanical strength have been investigated. However, the widespread use of solid electrolytes is limited by either their low ionic conductivity or narrow electrochemical window. Furthermore, the fabrication of cells with solid electrolytes encounters significant difficulties due to the substantial interfacial resistance between the electrodes and electrolytes. Consequently, there is currently no ideal solid electrolyte available to replace the commercially used liquid electrolyte [2].

Shear thickening electrolytes (STEs) are generally prepared by distributing solid inorganic fillers in liquid electrolyte, which exhibits liquid-like behavior under regular operating conditions, but demonstrates nonlinear shear thickening properties

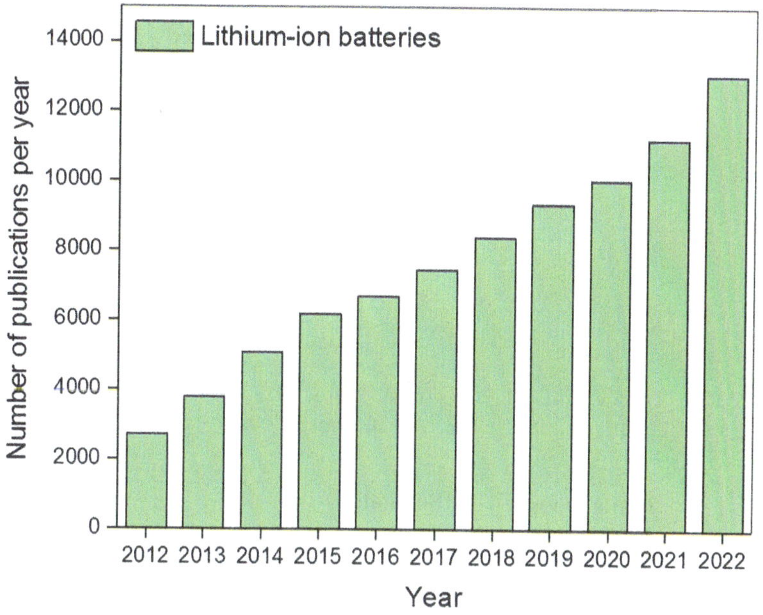

Fig. 4.2 The number of publications related to LIBs based on the Scopus database

when subjected to rapid external shear rates. When the shear thickening is triggered, the electrolyte displays quasi-solid characteristics, thereby enhancing the safety of the battery. This capability of intelligent electrolytes to switch between liquid and solid states mitigates the risk of short circuits and thermal runaway under mechanical loading. According to the literature, the design and development of such electrolytes offer a promising approach to have safe and high-performance LIBs [1]. STE is expected to function similarly to a conventional liquid electrolyte under normal conditions, exhibiting high ionic conductivity and compatibility with electrodes. Nevertheless, it possesses the unique ability to rapidly transition into a solid-like state when subjected to high shear forces, thereby effectively withstanding external impacts. In recent years, an increase in research in this research area has been observed. The growth in publications concerning the LIBs over the last decade is shown in Fig. 4.2.

4.2 Lithium-Ion Batteries

LIBs generally consist of cathodes (positive electrode), anodes (negative electrode), and a separator containing an electrolyte. Lithium is housed in the anode and cathode. The electrolyte facilitates the movement of positively charged lithium ions between the anode and cathode through the separator. As lithium ions migrate, a positive charge is generated at the current collector, inducing the release of electrons

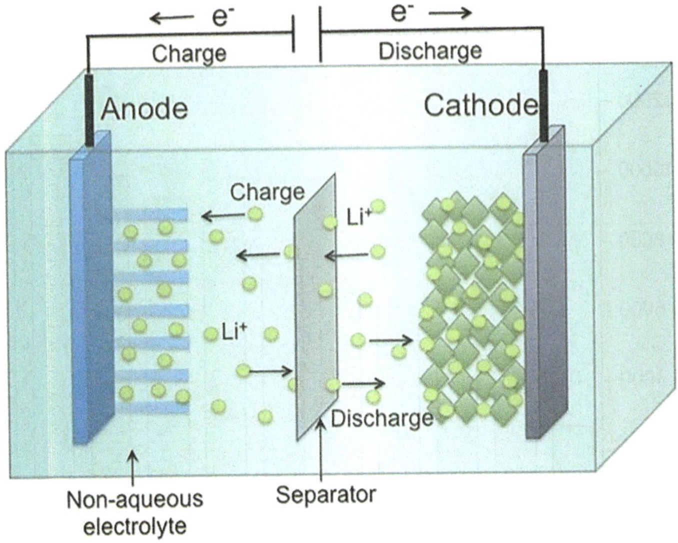

Fig. 4.3 The working principle of LIBs [4]

in the anode. Subsequently, the electrical current flows from the positive current collector to the negative one, passing through a powered device such as a computer or cell phone. Figure 4.3 shows the principles of LIBs. The anode releases lithium ions to the cathode as the battery discharges and produces an electric current, which causes a flow of electrons from one side to the other. The converse occurs when the gadget is plugged in, i.e., lithium ions are given out by the cathode and received by the anode. The energy density (Wh/kg) and power density (W/kg) are the two most common concepts of batteries. The former represents the quantity of energy that a battery can store in relation to its mass, while the latter denotes the amount of power that a battery can generate in relation to its mass.

The LIBs operate through the reversible movement of lithium ions between the cathode and anode materials. Hence, the performance of batteries is significantly affected by the inherent characteristics of electrode materials [3]. The cathode materials play a pivotal role in predominantly determining the overall performance of batteries compared to the anode materials. The separator serves to hinder the unrestricted movement of electrons within the battery. The electrochemical performance and safety of the batteries are influenced by electrolyte.

4.3 Shear Thickening Mechanism

The shear thickening phenomenon was initially observed in the 1960s. It was initially considered a problem in various industrial processes, such as mixing and coating. The shear thickening effect hindered fluid movement, resulting in uneven coating on material surfaces, blockage of crude oil in pipeline transportation, overload operation of high-speed agitators, and potential equipment damage. Due to its negative impact on industrial production, shear thickening received little attention at the time. However, in the mid-1990s, the development of a new type of shear thickening fluid (STF) through the use of nanotechnology sparked significant interest. This new STF has since been extensively studied and applied in various fields. The unique characteristics of STFs have been particularly useful in the development of smart materials and structures. One notable application of STFs is in the creation of thin, flexible, and affordable materials with excellent puncture, stab, and ballistic protective qualities compared to existing protective materials. By incorporating STFs into high-performance fabrics, researchers have been able to enhance the anti-impact properties of textile-based structures. This has led to investigations into the use of STFs for absorbing shock waves from severe wind or earthquake conditions, as well as integrating STFs into damper systems [5–8]. In addition to these applications, STFs have also shown potential in improving the vibration damping of structural components. By restricting the movement of joints such as elbows, ankles, knees, shoulders, and hips, STFs can prevent sudden accelerations and reduce the risk of injury. Overall, the discovery and development of shear thickening effects have opened up new possibilities for various industries and have sparked significant interest in the scientific community [9].

A schematic representation of shear thickening mechanism is shown in Fig. 4.4. In the equilibrium state, solid particles within the medium are evenly dispersed. When shear rate is below a certain threshold, STF exhibits a low viscosity. However, the particles experience a chaotic motion as shear rate increases, resulting in layered particle formations. With further increase in shear rate, hydroclusters begin to form, leading to the solid particles to agglomerate. Once the loading is removed, the suspensions turn back to the state of easy flow and continue to flow regularly. This shear thinning phenomenon is fully reversible [10, 11]. In the existing literature, researchers have proposed various explanations to elucidate the underlying mechanisms responsible for shear thickening behavior.

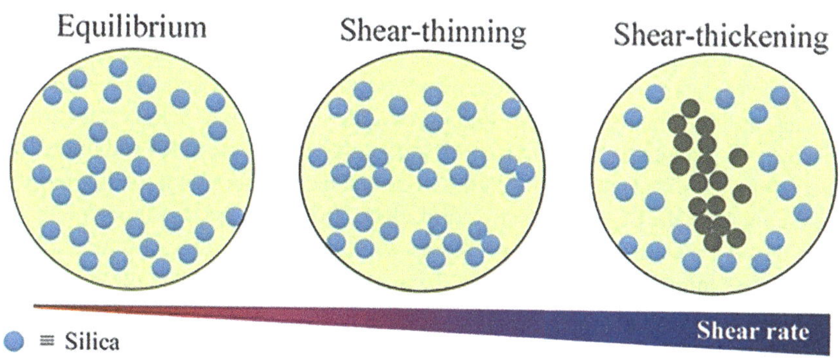

Fig. 4.4 Schematic representation of the shear thickening mechanism in LIB electrolytes

These behaviors are better explained by the order-disorder transition (ODT) [12–15] and the hydrocluster mechanism [16, 17] than by the other theories. The ODT describes how the breakdown of the solid particle formations, which results in an increase in drag forces on the particles, occurs when a shear rate is applied to the suspension. A micromechanical model of shear thickening as a flow-induced ODT was created by Hoffman [15], one of the initial investigators to extensively examine shear thickening, using this expertise. He postulated that the shear thickening tendency was caused by a change from a layered orientation, easily flowing condition to a disordered one. He proposed that a cluster of particles is subject to the electric double layer, van der Walls-London, and shear forces, which result in this transition. Furthermore, it was hypothesized that hydrodynamically generated forces act on the solid particles within the moving layers, pushing them outside of their layer and upsetting the flow's order. A sharp increase in viscosity happened when shear rate crossed a crucial point.

On the other hand, the hydrocluster mechanism is based on the interactions between the solid particles. It can be inferred from this mechanism that the shear thickening properties observed in concentrated colloidal mixtures is a result of hydrodynamic lubricating forces between the solid particles. The hydrodynamic pressure increases under a shear force application, and thereby colliding the particles suspended in the liquid. Because clusters form when there is a shear force, the drag forces between the particles are increased. The "hydrocluster" explanation is more thorough and precise than the other two microstructural descriptions for the shear thinning phenomenon.

4.4 Shear Thickening Electrolyte

LIBs have emerged as promising technologies for large-scale electrochemical energy-storage solutions owing to their exceptional energy densities, commendable charge retention capacity, and low toxicity. However, the practical applications of traditional LIBs have been significantly impeded due to safety concerns arising from the utilization of flammable organic-based electrolytes. The short circuits and electrolyte leakage generally occur when an external physical shock is applied to the batteries, leading to a fire or explosion. Solid electrolytes and ionic liquid electrolytes are the main topics of studies on safe lithium-ion battery electrolytes. However, in practical applications, both inorganic solid electrolytes and ionic liquid electrolytes confront major difficulties. For instance, ionic conductivity and electrode contact stability in inorganic solid electrolytes still need improvement. The ionic liquid electrolyte also has problems with low performance, a low number of lithium ions migration, and high cost. To overcome these obstacles, colloidal particles are suggested to be included in an inert dielectric solvent to create STEs. The utilization of STEs in battery systems can enhance the overall performance of the battery. They have strong ionic conductivity, excellent electrode compatibility, and combine the capabilities of solid and liquid electrolytes. In addition, the solidification of STEs at high shear rates, protects the batteries against short-circuiting and thermal runaway under physical shock. STEs show shear thickening behavior upon mechanical pressure or impact, absorbing the impact energy and having tolerance to failure. Therefore, multifunctional electrolytes with a good design can greatly safeguard batteries during the initial stages of battery thermal runaway [1].

Silica nanoparticles, as the most common type of filler, are widely utilized as fillers due to their ability to form hydroclusters through hydrogen bonding between their silanol groups. A significant loading of silica is required for sufficient contacts between the particles to produce hydroclusters in order to generate shear thickening effect. The appropriate range of fillers (silica nanoparticles) for the shear thickening effect has been estimated to be between 10 and 30 wt%. When fillers are added in such large quantities, the electrolyte becomes highly viscous, resulting in drawbacks, such as low conductivity and difficulty in injecting into the cell. It has been reported that the fillers with large aspect ratios tune the hydrodynamic forces and interparticle attractions to produce shear thickening effects at low content. Nevertheless, the viscosity remains high. These materials with high filler content have high mass densities, which lower the energy density in batteries. They face stability challenges due to the tendency of silica nanoparticles to aggregate and settle over time in frequently employed carbonate-based liquid electrolytes for LIBs. Consequently, to create impact-resistant LIBs, an effective STE with high conductivity, low viscosity, and increased stability is extremely desirable [18].

Although low viscosity can be easily attained by reducing the amount of filler content, it hinders the formation of hydroclusters, which are necessary for the production of a shear thickening effect. Hence, to tune the shear thickening, additives such as polymers, carbon nanotubes, graphene, etc. have been suggested to be introduced into the system. They effectively interconnect the particles, thereby promoting the formation of hydroclusters. For example, Chen et al. [18] prepared a new STE by incorporating a minor quantity of poly(ethylene oxide) (PEO) interacting with SiO_2 nanoparticles. They found that the utilization of PEO facilitates the hydrocluster formations and induces the shear thickening effect even at a low filler content of 2.2 wt%. The developed STE also exhibits a remarkable ionic conductivity of approximately 5.1 mS/cm with outstanding long-term stability over one month. Figure 4.5 shows the interaction of silica nanoparticles and PEO and its microstructure evolution under shear rates. The hydroxyl (-OH) groups on the PEO chain and the silanol groups on silica form a hydrogen bond, which causes the PEO to be adsorbed onto silica particles in the suspension. Low shear rates cause the polymer chain, which is in its naturally coiled state, to extend and electrostatic interactions to weaken, which increases the anisotropy of suspension and lowers viscosity. At low shear rates, the repulsive forces between particles such as PEO and PEO, silica particles and silica particles, and PEO and silica particles are dominant. After the critical shear rate, the repulsive force between "particles" is broken, allowing hydrodynamic forces to prevail over the repulsive force. This enables elongated polymer chains to connect with other silica particles, resulting in the formation of extensive three-dimensional interconnected aggregates, commonly referred to as shear-induced clusters. The presence of these hydroclusters hinders the flow of fluid and viscosity is significantly raised. The higher shear stress forced these interconnected aggregates to break down, leading to a reduction in viscosity.

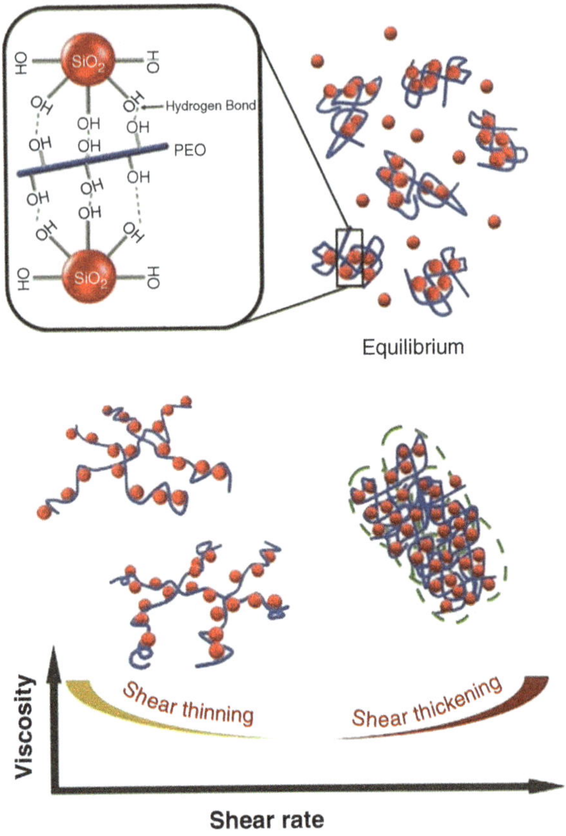

Fig. 4.5 Schematic illustration of the interaction of silica nanoparticles and PEO in STE and microstructure evolution for reversible shear thinning and shear thickening under shear rates [18]

Chen et al. [18] found that by increasing the silica content from 2 to 4 wt%, shear thickening effects were more pronounced and the critical shear rate decreased at the same time. This could be explained by the larger, more powerful hydroclusters that were created as a result of more silica particles adhering to the polymer chains. The shear thickening magnitude was largest in the suspension made up of 4 wt% SiO_2 and 0.2 wt% PEO, which also had the largest peak viscosity of 1.68 Pa at the shear rate of 61.9 s^{-1}. Investigating the effect of PEO concentration on the shear thickening properties of STE containing 4% silica showed an inverse effect of increasing PEO concentration on the magnitude of shear thickening. Notably, the shear thickening behavior became insignificant at a PEO content of 0.8 wt%, attributing to the blocking of the required adsorption sites for the formation of hydroclusters through bridging silica particles at an excess amount of PEO.

The effect of the aspect ratio of anisotropic silica nanorods on STEs has been investigated by Ye et al. [19] (Fig. 4.6a). They found that the critical volume fraction for shear thickening lowers by increasing aspect ratio. Figure 4.6b shows the rheogram of STE having silica nanorods with various volume fractions and aspect ratios. Incorporation of nanorods with an aspect ratio of 24 at a volume fraction of 0.146 leads to discontinuous shear thickening where solidification occurs. Interparticle attractions give rise to hysteresis in the critical shear stress during shear thickening.

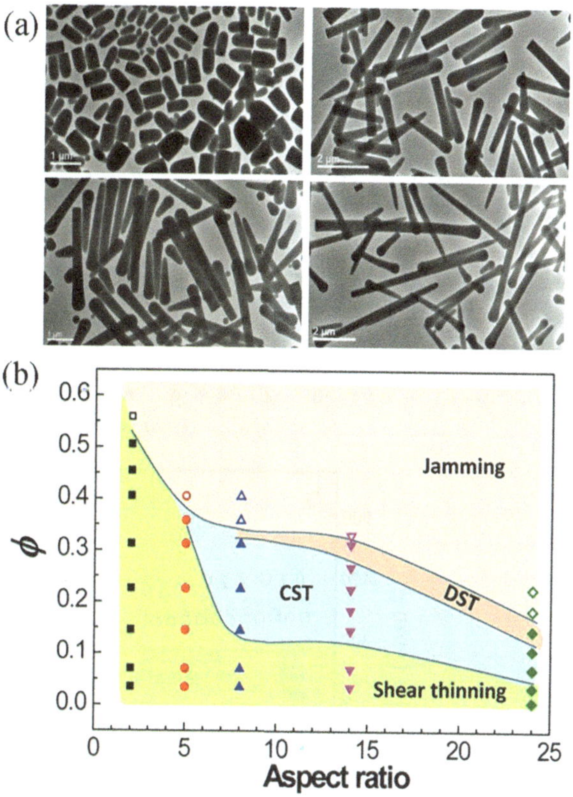

Fig. 4.6 (**a**) TEM images of silica nanorods with different aspect ratios, (**b**) rheogram of STE containing silica nanorods with different aspect ratios and volume fractions [19]

This is due to the contact between the solvation layer of the nanorod and the electrolyte. Additionally, increasing the aspect ratio leads to a decrease in interparticle attractions. This is likely associated with the increased surface area of the silica nanorods. As a result, hydrodynamic interactions, frictional contacts, interparticle attractions, and particle aspect ratio all act together to cause shear thickening.

Ding et al. [20] also developed STE by incorporating fumed silica nanoparticles into an electrolyte (1 M LiFP6 in EC/DMC) which is generally used in commercial LIBs. They found that the STE containing 10.7 wt% silica particles processes the largest shear thickening behavior. Moreover, it had higher ionic conductivities (1.93×10^{-2} S/cm) than the commercial liquid electrolytes (2×10^{-3} S/cm). Figure 4.7a shows the variation of ionic conductivity of prepared STE as a function of fumed silica content. The cell test results (Fig. 4.7b) also showed that the reversible capacity of the cell using STE was marginally higher than that of the traditional electrolyte. Moreover, a comparison of Nyquist plots of the cell before and after the impact test (Fig. 4.7c) showed only a mild change in electrochemical impedance, indicating that the developed STE did not negatively affect the electrochemical performance. Their results confirmed that the STE demonstrates the capability to improve battery system safety with excellent resistance to crushing while maintaining the fundamental electrochemical performance.

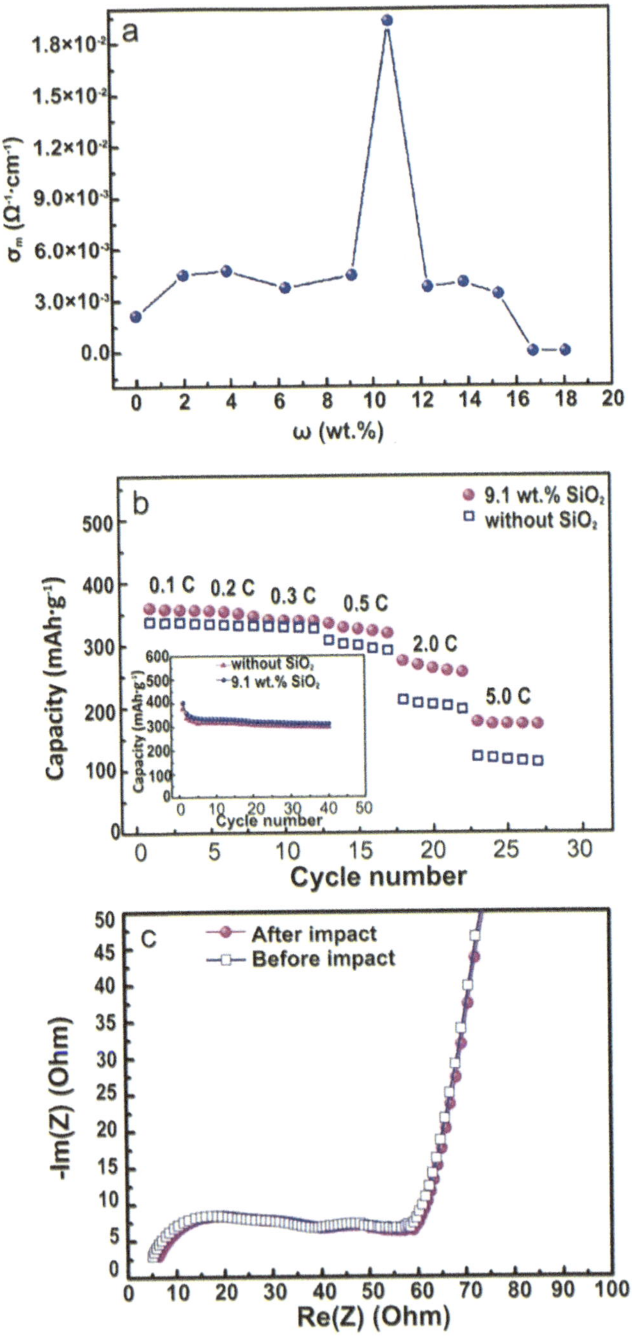

Fig. 4.7 (**a**) Variation of ionic conductivity of STE ($SiO_2/LiPF_6$ in EC/DMC) as a function of fumed silica content at room temperature. (**b**) Rate performance of graphite electrode with STE (SiO_2 weight fraction of 9.1%). (**c**) Electrochemical impedance spectra (EIS) of the cell before and after impact test [20]

4.5 Mechanical Performance of Batteries with STE

To study the possibility of enhancing impact resistance of LIBs reinforced with STE, a ballistic impact test was carried out by Ye et al. [19]. They prepared target consisting of soft (Level II) and hard (Level III+) armor plates, a battery with STE, pressure-sensitive film, and polymer clay back-face. The obtained results revealed the significant difference between the LIB with and without STE, confirming the pivotal role of STE in providing the protection of LIBs. Figure 4.8 shows the ballistic test results of soft and hard armor. Using the STE in batteries for both Level II and Level III+ armor, the depth of impact was found to be shallower. Moreover, the pressure-sensitive film showed a clear difference in the impact force distribution. When the STE was used, there was a noticeably decreased presence of pink coloring in the films.

Investigating the anti-crushing capability of batteries with STE by Ding et al. [20] showed excellent resistance to crushing, improving the mechanical safety of batteries. The protective mechanism of STE is schematically represented in Fig. 4.9. The electrolyte condition before the impact is shown in Fig. 4.9b, where the particles are slightly attracted to one another and the fumed silica particles improve ionic conductivity. The particles are lubricated by the low-viscosity colloidal suspension, which causes them to disperse uniformly throughout the liquid without forming clumps or sinking to the bottom. When an impact force is given to an electrolyte (Fig. 4.9c), it overcomes the repulsive interactions between the particles to encourage particle clustering. As a result, the electrolyte takes on a solid-like form and becomes less penetrable. The particles oppose one another once more and the hydroclusters disintegrate as the impact energy dissipates (Fig. 4.9d), returning the pseudo-solid substance to its initial liquid state. On the contrary, in the absence of a shear thickening effect, the battery system is susceptible to significant deformation and a quick temperature rise during impact, which could potentially trigger an ignition in the solvents.

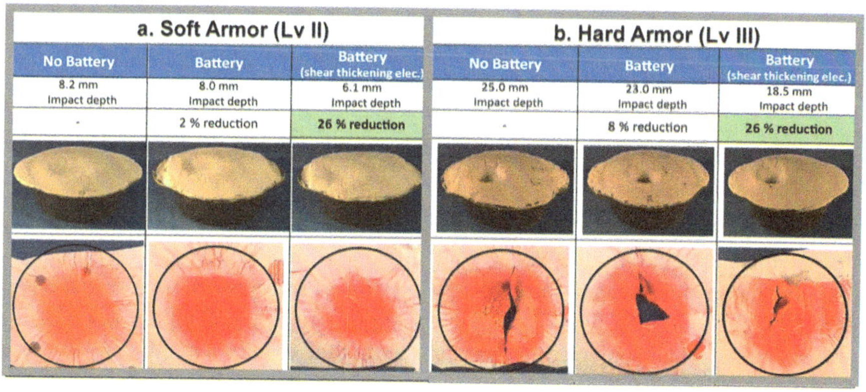

Fig. 4.8 The results of the ballistic test of a target consisting battery with STE, (**a**) soft and (**b**) hard armor [19]

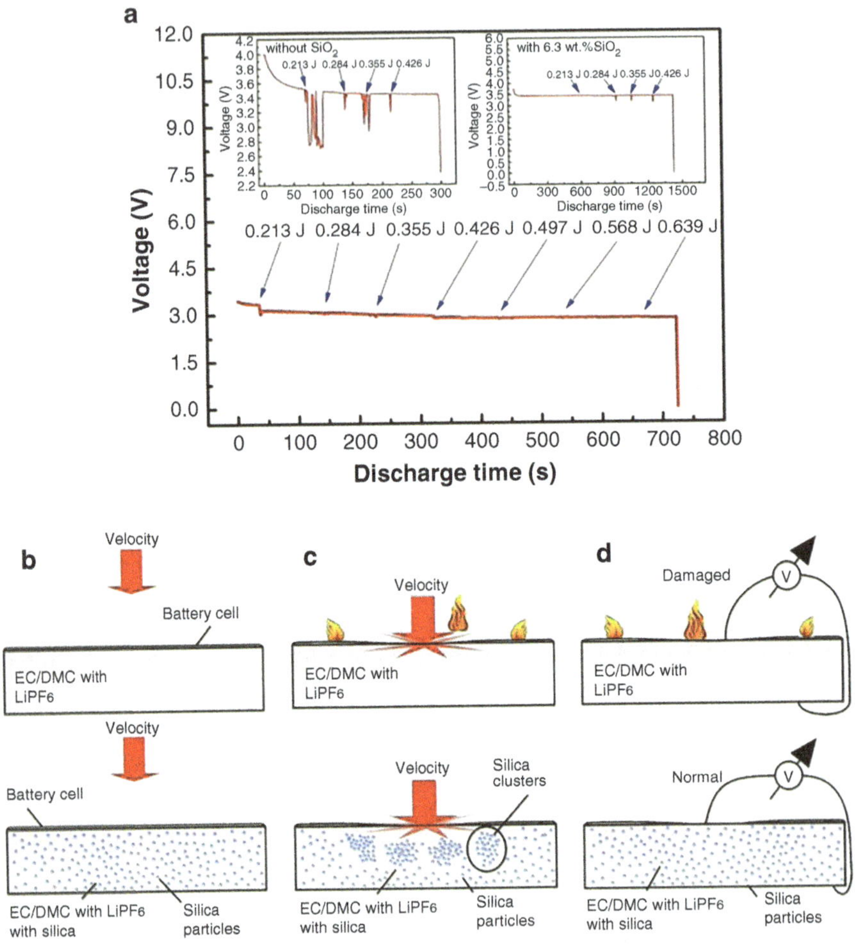

Fig. 4.9 Schematic illustration of the protective mechanism of STE in battery systems. Top: The conventional electrolyte used in batteries is susceptible to internal shorting due to mechanical impact, which can result in hazardous incidents such as fires and explosions. Bottom: The STE under pressure or impact has shown remarkable resistance to crushing [20]

4.6 Summary

In this chapter, the recent progress on LIBs based on STEs has been reviewed. The discussion encompasses the underlying principle, electrochemical performance, and practical applications of these battery systems. It is important to note that STEs exist as non-Newtonian fluids, with their normal state being a slightly viscous liquid. However, when subjected to external shock, they exhibit a significant increase in shear storage modulus and other rheological properties. Upon removal of the impact loading, the system reverts back to its original liquid state. The shear thickening effect in the electrolyte system has been found to be influenced by factors such as the size and mass fraction of suspended particles, dispersion medium, and temperature. The shear thickening behavior plays an important role in protecting the battery system from mechanical loadings, such as impacts or shocks. Electrochemical and ballistic tests have demonstrated the capability of STEs to convert from liquid to solid phases or vice versa. This unique characteristic is beneficial for the safety of the battery system under certain conditions and combines the benefits of both solid and liquid electrolytes, including the stable electrochemical characteristics of liquid electrolytes, and the improved safety performance of solid electrolytes. The utilization of STEs will also reduce the requirement for additional weight or volume to achieve sufficient protection. Although some studies have focused on STEs, the practical usage of high-performance LIBs is impeded by a range of challenges in the STE development, which is still in the early stages. For instance, impact resistance performance, ionic conductivity, electrochemical properties, and shear thickening mechanism are among several key criteria to evaluate the performance of STEs. Moreover, achieving a shear thickening effect with low additive concentration and minimizing energy density loss should be further explored. It is anticipated that forthcoming studies will focus on developing more efficient strategies to enhance the performance of LIB battery systems with STEs.

References

1. Huang, Q., Liang, X., Liu, B., Deng, H.: Research progress of shear thickening electrolyte based on liquid–solid conversion mechanism. Batteries. **9**, 1–12 (2023). https://doi.org/10.3390/batteries9070384
2. Liu, K., Cheng, C.F., Zhou, L., Zou, F., Liang, W., Wang, M., Zhu, Y.: A shear thickening fluid based impact resistant electrolyte for safe Li-ion batteries. J. Power Sources. **423**, 297–304 (2019). https://doi.org/10.1016/j.jpowsour.2019.03.056
3. Xu, G.L., Wang, Q., Fang, J.C., Xu, Y.F., LiPhD, J.T., HuangPhD Prof, L., Sun, S.G.: Tuning the structure and property of nanostructured cathode materials of lithium ion and lithium sulfur batteries. J. Mater. Chem. A. **2**, 19941–19962 (2014). https://doi.org/10.1039/c4ta03823a
4. Roy, P., Srivastava, S.K.: Nanostructured anode materials for lithium ion batteries. J. Mater. Chem. A. **3**, 2454–2484 (2015). https://doi.org/10.1039/C4TA04980B
5. Hasanzadeh, M., Mottaghitalab, V., Babaei, H., Rezaei, M.: The influence of carbon nanotubes on quasi-static puncture resistance and yarn pull-out behavior of shear-thickening fluids (STFs) impregnated woven fabrics. Compos. Part A Appl. Sci. Manuf. **88**, 263–271 (2016). https://doi.org/10.1016/j.compositesa.2016.06.006
6. Mirrahimi, A.H., Hasanzadeh, M., Mottaghitalab, V., Sharma, P.: Numerical modelling of ballistic impact on HMPP woven fabric impregnated with shear-thickening fluids. Procedia Eng. **173**, 73–76 (2017). https://doi.org/10.1016/j.proeng.2016.12.031
7. Hasanzadeh, M., Mottaghitalab, V.: The role of shear-thickening fluids (STFs) in ballistic and stab-resistance improvement of flexible armor. J. Mater. Eng. Perform. **23**, 1182–1196 (2014). https://doi.org/10.1007/s11665-014-0870-6
8. Hasanzadeh, M., Mottaghitalab, V., Sharma, P.: Puncture resistance enhancement of woven fabrics using concentrated nanosilica suspension. Procedia Eng. **173**, 1494–1498 (2017). https://doi.org/10.1016/j.proeng.2016.12.228

9. Sheikhi, M.R., Hasanzadeh, M., Gürgen, S.: The role of conductive fillers on the rheological behavior and electrical conductivity of multi-functional shear thickening fluids (M-STFs). Adv. Powder Technol. **34**, 104086 (2023). https://doi.org/10.1016/j.apt.2023.104086
10. Hasanzadeh, M., Mottaghitalab, V.: Tuning of the rheological properties of concentrated silica suspensions using carbon nanotubes. Rheol. Acta. **55**, 759–766 (2016). https://doi.org/10.1007/s00397-016-0950-7
11. Hasanzadeh, M., Mottaghitalab, V., Rezaei, M.: Rheological and viscoelastic behavior of concentrated colloidal suspensions of silica nanoparticles: a response surface methodology approach. Adv. Powder Technol. **26**, 1570–1577 (2015). https://doi.org/10.1016/j.apt.2015.08.011
12. Hoffman, R.L.: Explanations for the cause of shear thickening in concentrated colloidal suspensions. J. Rheol. (N. Y. N. Y). 42 (1998) 111. https://doi.org/10.1122/1.550884
13. Boersma, W.H., Laven, J., Stein, H.N.: Viscoelastic properties of concentrated shear-thickening dispersions. J. Colloid Interface Sci. **149**, 10–22 (1992). https://doi.org/10.1016/0021-9797(92)90385-Y
14. Laun, H.M., Bung, R., Hess, S., Loose, W., Hess, O., Hahn, K., Hädicke, E., Hingmann, R., Schmidt, F., Lindner, P.: Rheological and small angle neutron scattering investigation of shear-induced particle structures of concentrated polymer dispersions submitted to plane Poiseuille and Couette flowa), J. Rheol. (N. Y. N. Y). 36 (1998) 743. https://doi.org/10.1122/1.550314
15. Hoffman, R.L.: Discontinuous and dilatant viscosity behavior in concentrated suspensions. II. Theory and experimental tests. J. Colloid Interface Sci. **46**, 491–506 (1974). https://doi.org/10.1016/0021-9797(74)90059-9
16. Bender, J.W., Wagner, N.J.: Optical measurement of the contributions of colloidal forces to the rheology of concentrated suspensions. J. Colloid Interface Sci. **172**, 171–184 (1995). https://doi.org/10.1006/JCIS.1995.1240
17. Bender, J., Wagner, N.J.: Reversible shear thickening in monodisperse and bidisperse colloidal dispersions, J. Rheol. (N. Y. N. Y). 40 (1998) 899. https://doi.org/10.1122/1.550767
18. Chen, Z., Chao, Y., Sayyar, S., Tian, T., Wang, K., Xu, Y., Wallace, G., Ding, J., Wang, C.: Polyethylene oxide (PEO) provides bridges to silica nanoparticles to form a shear thickening electrolyte for high performance impact resistant Lithium-ion batteries. Adv. Sci. **2302844**, 1–8 (2023). https://doi.org/10.1002/advs.202302844
19. Ye, Y., Xiao, H., Reaves, K., McCulloch, B., Mike, J.F., Lutkenhaus, J.L.: Effect of Nanorod aspect ratio on shear thickening electrolytes for safety-enhanced batteries. ACS Appl. Nano Mater. **1**, 2774–2784 (2018). https://doi.org/10.1021/acsanm.8b00457
20. Ding, J., Tian, T., Meng, Q., Guo, Z., Li, W., Zhang, P., Ciacchi, F.T., Huang, J., Yang, W.: Smart multifunctional fluids for lithium ion batteries: enhanced rate performance and intrinsic mechanical protection. Sci. Rep. **3**, 1–7 (2013). https://doi.org/10.1038/srep02485

Chapter 5
Shear Thickening Fluid-Based Vibration Damping Applications

Ömer Yay, Seyid Fehmi Diltemiz, Melih Cemal Kuşhan, and Selim Gürgen

5.1 Introduction

Vibration is a ubiquitous phenomenon in various engineering systems, ranging from aircraft and automobiles to buildings and industrial machinery. While controlled vibrations are essential for the efficient functioning of many devices, uncontrolled vibrations can lead to performance degradation, structural fatigue, and even catastrophic failures [1]. Thus, effective vibration damping methods are crucial to ensure the longevity, reliability, and safety of these systems.

In engineering applications, vibration damping refers to the attenuation of undesirable mechanical vibrations by dissipating energy. This energy dissipation minimizes the transmission of vibrations from one part of a system to another, resulting in reduced noise, enhanced comfort, and extended operational lifespan. Vibration damping finds applications in an array of industries, including aerospace, automotive, civil engineering, and sports equipment manufacturing. Traditional methods of vibration damping, such as passive systems using rubber mounts, elastomers, and viscous fluids, have been extensively employed to mitigate vibrations [2]. While these methods are effective to some extent, they often exhibit limitations. Passive dampers are limited by their fixed damping coefficients and lack adaptability to varying vibrational conditions. Moreover, these methods may not effectively attenuate vibrations across a wide frequency range, leading to resonances and inadequate damping performance.

The demand for improved vibration damping performance has spurred the exploration of advanced materials and technologies. One such promising avenue is the

Ö. Yay (✉)
Department of Aeronautical Engineering, Gebze Technical University, Gebze, Türkiye
e-mail: omeryay@gtu.edu.tr

S. F. Diltemiz · M. C. Kuşhan · S. Gürgen
Department of Aeronautical Engineering, Eskişehir Osmangazi University, Eskişehir, Türkiye

S. Gürgen (ed.), *Smart Systems with Shear Thickening Fluid*,
https://doi.org/10.1007/978-3-031-53570-3_5

utilization of shear thickening fluids (STFs) as damping agents. STFs exhibit a unique rheological behavior characterized by their ability to transition from a low-viscosity state to a highly viscous or even solid-like state under mechanical shear [3, 4]. This distinctive behavior makes STFs particularly attractive for vibration damping applications [5–7].

5.2 STFs in Vibration Damping

STFs have taken significant attention for their intriguing and versatile rheological behavior, which makes them well-suited for various applications, including vibration damping. STFs are colloidal suspensions comprising solid particles dispersed in a liquid medium. Their unique behavior arises from the interactions between these particles under the influence of shear stress. As shear rate increases, the particles form transient particle networks that obstruct the flow of the suspending liquid. This network formation results in a rapid increase in viscosity and resistance to further deformation, leading to shear thickening. The key to shear thickening mechanism in STFs lies in the interparticle interactions [8]. Under low shear conditions, particles are dispersed randomly and experience minimal interaction. However, as shear stress is applied, particles come into contact and form temporary contacts or physical bonds. These bonds create interconnected structures, known as shear-induced particle networks, which hinder the flow of the fluid [9].

Several factors influence the shear thickening behavior in STFs, and understanding these factors is crucial for tailoring STF formulations for specific applications. Particle properties, such as size, shape, and surface chemistry, play a significant role in determining the strength of interparticle interactions. Additionally, the concentration of particles in the suspension and the properties of the suspending medium, such as viscosity and rheological behavior, impact the onset and intensity of shear thickening [10, 11]. The rheological behavior of STFs exhibits distinctive characteristics that make them suitable for vibration damping applications. At low shear rates, STFs exhibit low viscosity magnitudes. However, as the shear rate increases, the viscosity undergoes a sudden and significant increase due to shear thickening mechanism. This increase in viscosity is often accompanied by a sharp increase in yield stress, indicating the transition from a liquid-like to a solid-like behavior. The ability of STFs to transition from a low-viscosity state to a high-viscosity or solid-like state under shear stress is a fundamental property that can be harnessed for vibration damping applications [12]. This unique behavior allows STFs to absorb and dissipate energy from mechanical vibrations, making them a promising alternative to conventional damping materials.

The shear thickening phenomenon in STFs is directly linked to shear rate. Higher shear rates lead to more pronounced shear-induced particle interactions and stronger particle networks, resulting in more substantial shear thickening. Moreover, temperature can also influence the shear thickening behavior by affecting the mobility and interactions of particles within the suspension [13, 14]. Understanding the effects of shear rate and temperature is crucial for tailoring STF formulations for specific applications. On the other hand, the material properties of STFs play a crucial role in determining their effectiveness for vibration damping. Higher viscosity and yield stress contribute to increased energy dissipation and damping capabilities. The properties of the solid particles dispersed within STFs are integral to their rheological behavior. Particle size, shape, surface roughness, and chemical composition influence the strength of interparticle interactions and network formation. The choice of suspension medium, which can be a Newtonian or non-Newtonian fluid, also impacts the overall rheological behavior of the STF. Careful selection of particle characteristics and suspension medium is essential for tailoring STF properties to meet specific damping requirements. While the unique shear thickening behavior of STFs offers significant advantages for vibration damping applications, certain challenges and considerations have to be addressed. Stability issues, especially in long-term applications, and the potential for material agglomeration under high shear conditions are among the challenges that researchers and engineers have to tackle. Additionally, the compatibility of STFs with other materials and environmental factors requires careful evaluation.

The viscoelastic behavior of STFs is central to their effectiveness in damping vibrations [15]. As vibrations are imparted to the material, STFs undergo stress relaxation and creep, absorbing and dissipating energy in the process. This viscoelastic response is due to the temporary bonding between particles within the shear-induced networks. The energy absorbed during these processes is converted into heat, resulting in effective vibration damping across a wide frequency range. Another key mechanism in STF-based vibration damping is particle rearrangement. Under the influence of mechanical stress, particles within the STF migrate and rearrange, leading to changes in the microstructure of the material. This microstructural evolution creates additional pathways for energy dissipation through friction and interparticle interactions. As a result, mechanical energy is converted into heat, contributing to the overall damping effect. STFs exhibit an increase in viscosity and yield stress under shear, which can lead to significant frictional forces within the material. These internal frictional forces contribute to the dissipation of vibrational energy, especially in regions where high shear stress occurs. The frictional damping mechanism is particularly effective in dampening high-frequency vibrations, complementing the viscoelastic damping mechanism. The ability of STFs to simultaneously utilize viscoelastic and frictional damping mechanisms allows them to damp vibrations across a broad frequency spectrum. This frequency-dependent damping behavior is a valuable advantage over conventional damping materials, which often struggle to provide efficient damping across a wide range of frequencies. STFs can effectively attenuate vibrations at both low and high frequencies, making them versatile solutions for diverse engineering applications. The synergistic interplay enables STFs to exhibit superior damping performance compared to single-mechanism damping materials. This comprehensive approach ensures effective energy dissipation across various vibrational conditions.

The aerospace industry demands lightweight and efficient solutions to mitigate vibrations in aircraft and spacecraft. STFs offer an attractive alternative to conventional damping materials in reducing structural vibrations, noise, and undesired oscillations [16]. Passenger comfort, fuel efficiency, and lifespan of critical components can be enhanced by integrating STF-based damping systems into aircraft components and interior structures. Similarly, STF-based vibration damping solutions hold promise for enhancing vehicle comfort, safety, and performance in the automotive engineering. STFs can be incorporated into suspension systems, chassis components, and interior panels to dampen road-induced vibrations and reduce noise. This leads to improved driving experience, decreased driver fatigue, and prolonged vehicle durability. Another application area is structural engineering for STFs [17, 18]. Structural vibrations in buildings, bridges, and other infrastructure can lead to long-term damage and reduced lifespan. By introducing STF-based damping systems into these structures, vibrations caused by external forces, such as wind, earthquakes, and dynamic loads can be effectively mitigated. STFs offer a non-intrusive approach to reducing vibrations in structures, ensuring both safety and functionality. Alternatively, STFs are making their mark in the sports equipment, offering enhanced performance and safety across various athletic disciplines [19]. From tennis rackets to ski bindings, STFs can be integrated into equipment designs to absorb and dissipate impact energy, providing athletes with improved control, reduced risk of injury, and increased performance potential. Beyond aerospace, automotive, structural engineering, and sports equipment, STFs are finding applications in diverse sectors. The medical field, for instance, could benefit from STF-based solutions to enhance the performance of prosthetics and medical devices. Additionally, industrial machinery and robotics could employ STFs to reduce vibrations and enhance precision in manufacturing processes. Successful integration of STF-based damping solutions requires a tailored approach to each application. Factors such as material compatibility, environmental conditions, and the specific vibrational challenges posed by each industry have to be considered. Customizing STF formulations and integration methods ensures optimal performance and addresses the unique demands of different engineering contexts.

5.3 Challenges in STF-Based Vibration Damping Applications

The stability of STF-based damping systems over extended periods remains a concern. Particle settling, agglomeration, and changes in rheological behavior over time can impact their damping efficiency. Temperature is another key factor on STF rheology. Viscosity profile of STF is heavily dependent on temperature. Researchers are actively working on understanding and mitigating these stability challenges to ensure that STFs deliver consistent and reliable performance throughout their operational lifespan [20–22]. Another challenge is including STF in conventional engineering systems. Integrating STFs with existing materials and systems requires careful consideration of material compatibility. STF interactions with surrounding materials, such as metals, polymers, and composites, can affect their damping efficiency and overall performance. Finding suitable combinations and developing methods to ensure compatibility without compromising STF properties is a critical task. The sustainability of STFs and their environmental impact have to be addressed also. Evaluating the life cycle of STF production, usage, and disposal is crucial to ensure that their benefits outweigh potential negative effects. Researchers and manufacturers are exploring environmentally friendly formulations and production processes to minimize the ecological footprint of STF-based damping systems [23–25]. The economic feasibility of STF-based damping solutions is another factor to consider. Manufacturing STFs, integrating them into engineering systems, and maintaining their performance can incur costs. Striking a balance between the benefits gained from reduced vibrations, improved performance, and extended lifespan, and the associated costs is essential for widespread adoption.

The field of material science offers exciting opportunities for enhancing STF properties. Innovations in particle design, surface modification, and suspension mediums can lead to improved damping efficiency, stability, and compatibility. Researchers are exploring novel materials and optimizing particle characteristics to tailor STFs for specific applications [26–28]. Future directions in STF-based vibration damping include the development of smart and adaptive damping systems. By integrating sensors and control mechanisms, researchers can create damping solutions that dynamically adjust their properties in response to changing vibrational conditions. These systems have the potential to optimize damping efficiency and adapt to varying operational scenarios. On the other hand, nanotechnology offers avenues for enhancing STF properties at the nanoscale. The incorporation of nanoparticles into STF formulations can fine-tune their rheological behavior and damping capabilities. Furthermore, the concept of multifunctionality, where STFs serve multiple roles beyond damping, is gaining interest [29]. For instance, STFs with self-healing properties could provide extended lifespan and improved durability.

5.4 A Case Study: STF-Based Vibration Isolation Foams

5.4.1 Experimental Details

In this study, fumed silica and polyethylene glycol (PEG)-based STF was integrated into polyester foams to have vibration damping pads with enhanced vibration damping properties. Polyester foams with an areal density of 1.10 kg/m^2 were obtained from Espol Inc. in 10 mm thickness. Fumed silica with an average particle size of 20 nm was supplied from Evonik. PEG with a molecular weight of 400 g/mol was supplied from Sigma Aldrich. STF was fabricated by distributing 60 wt% of fumed silica particles in PEG medium. A high-speed homogenizer was used for 1 h to effectively distribute the fumed silica particles in the suspension. The mixture was

Table 5.1 Details of specimens in vibration tests

Specimen	Details
NPF	Neat polyester foam
PF/STF	STF impregnated polyester foam
PF/STF/CNT	STF/CNT impregnated polyester foam

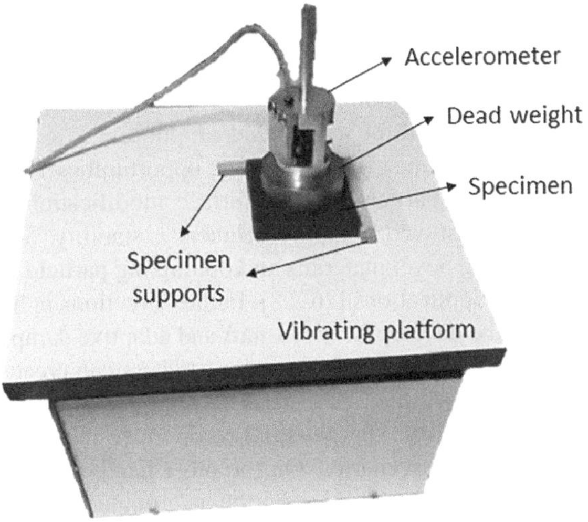

Fig. 5.1 Vibration test system

rested for 1 day to remove the air bubbles after homogenization process. In addition to the pristine STF, another STF mixture was produced by distributing 0.5 wt% of carbon nanotube (CNT) fillers having the size; D × L: 6–9 nm × 5 μm. An MCR 301 system by Anton Paar was used to obtain the rheological flow curves of the STF samples. The measurements were carried out with 25 mm diameter parallel plates keeping the sample gap at 1 mm. Shear rate was increased from 0 to 1000 s^{-1} during the measurements.

Polyester foams were impregnated with pristine STF and CNT included STF. At this stage, 50 × 50 mm^2 foam layers were immersed in the suspensions and then rested at room conditions to remove the excessive STF. Table 5.1 gives the details of specimens investigated in vibration tests. After one-day resting, the polyester foams were tested in a forced vibration test setup as shown in Fig. 5.1. The specimens were leaned on the supports on the vibrating platform. A dead weight of 1 kg attached to an accelerometer was placed on the specimens. The vibrating platform, which was mounted to the base plate with four springs, was excited by an eccentric motor. The accelerometer collected the vertical oscillations for the specimens.

5.4.2 *Results and Discussion*

5.4.2.1 STF Rheology

Figure 5.2 shows the rheological flow curves for the pristine and CNT included STF samples. As given in the graph, both suspensions show shear thickening onset at about 10 s^{-1}. Beyond this shear rate, which is known as critical shear rate, suspension viscosities exhibit sudden increase. The viscosity jump stems from the interactions between the silica particles in the liquid medium. As shear rate increases in the medium, hydrodynamic forces grow stronger on the silica particles. For this reason, interparticle distances get smaller, leading to formation of particle aggregations in the flow field. These particle groups are called hydroclusters, which are responsible for hindering the flow and consequently increasing the suspension viscosity. According to previous studies [30], hydrodynamic interactions cause the particles to come into physical contact with each other. Hence, contacted microstructures are obtained during shear thickening process. Although mild shear thickening behavior is due to the hydrodynamic interactions, contacted microstructure is responsible for strong thickening effect. Particle contacts produce chain-like pathways for the developed stresses in the suspensions during shear thickening process. The suspensions bear the stresses up to the peak viscosity point while the contacted microstructure shows a structural breakdown beyond the shear rates corresponding to the peak viscosities. Thus, the suspensions show obvious viscosity drops immediately after the peak points as shown in the charts.

It is possible to mention that CNT fillers enhance the suspension viscosity due to the increased amount of solid phase in the liquid medium. On the other hand, peak viscosity is increased in the CNT included STF. Hence, it can be stated that the addition of CNT reinforces the shear thickening behavior. CNT fillers are situated within the silica particle groups, enhancing the entanglement during the hydroclustering

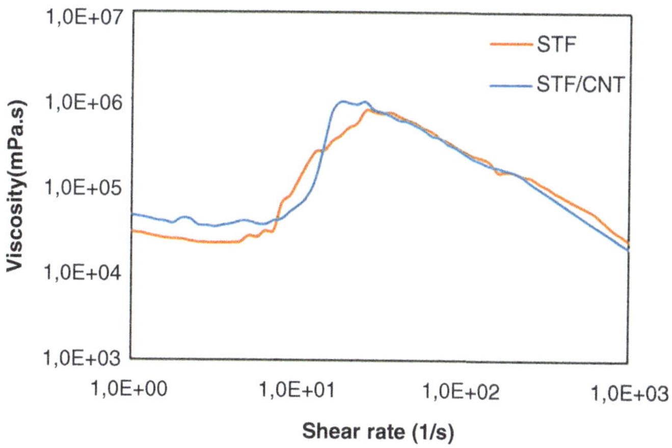

Fig. 5.2 Rheological flow curves for the STF samples

process. For this reason, hydroclustering extends to larger volumes within the liquid medium so that viscosity increase gets stronger. Another effect by CNT fillers is bonding interactions between the silica particles, carrier liquid, and CNT. Silanol groups on the silica particles produce hydrogen bonds with hydroxyl groups in PEG. Similarly, with the presence of CNT fillers, hydroxyl/carboxyl groups on these fillers can produce hydrogen bonds with the silica particles and PEG. By the way, bonding with PEG is generally stronger than that with silica particles. The flow is also hindered by these bonds and thereby leading to an additional contribution to the viscosity increase in the suspension. Consequently, shear thickening is supported by CNT fillers by this way [31].

5.4.2.2 Vibration Damping of STF Impregnated Foams

Figure 5.3 shows the accelerations in vertical direction for the specimens. According to the results, accelerations of the dead weight are in the range of 1.30 g by using the neat polyester foam. On the other hand, the acceleration ranges are about 0.80 g and 0.65 g for the cases with STF and STF/CNT impregnated polyester foams, respectively. It is clearly seen that accelerations on the dead weight are lowered by integrating STF into the polyester foam. Furthermore, the reduction in the accelerations is higher by using CNT included STF in the foam impregnation. The reduction is about 38% with the STF impregnated foam while it is about 50% with the CNT included STF-impregnated foam in comparison to the neat polyester foam.

STF effect is obvious in vibration damping. When the STF impregnated foam is excited, hydrodynamic effects on the silica particles grow stronger. Flowability in the suspension gradually decreases due to the effects of hydrodynamic interactions,

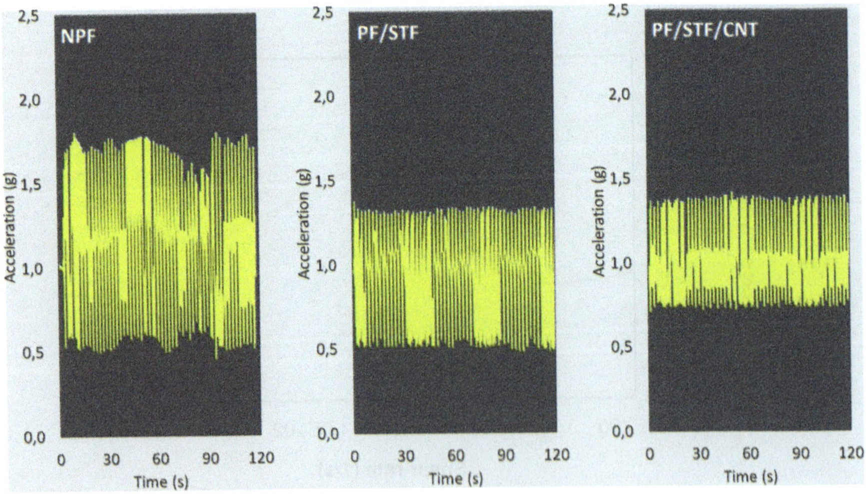

Fig. 5.3 Accelerations in vertical direction for the specimens

as the accumulated silica particles impede the fluid's ability to flow. Contacted silica particles during shear thickening process create pathways for energy dissipation through friction and interparticle interactions. The contacted microstructure provides an energy-conversion mechanism from mechanical vibration energy to heat energy, and thereby leading to a damping effect in the foam. Shear thickening behavior provides an energy attenuation mechanism for the energy input to the structure from the vibrations. STF in the polyester foams shows frictional interactions, and therefore, contributes to the dissipation of vibrational energy. The frictional interactions are pronounced especially at high shear stress areas in the polyester foams. This interaction also provides a vibration damping process in the structures.

Another important point is that STF shows different magnitudes of viscosities based on the shear rates during vibrational loadings. For this reason, an adaptive vibration damping behavior is obtained with STF. According to previous studies [32, 33], shear thickening process takes place even at quasi-static conditions that corresponds to very small shear rates. At low shear rates, shear thickening behavior is quite mild, so that moderate viscosity levels are observed in the suspensions. However, drastic viscosity jumps are visible in the suspensions at high level shear rates. Vibration damping systems efficiently benefit from the viscoelastic behavior of STF by this way. Apart from non-Newtonian rheology, Newtonian fluids are also effectively used for attenuating vibrational loadings in engineering structures, such as bridges, wind turbines, and automobiles [12]. Viscous nature of Newtonian fluids shows mild sloshing under vibrational loadings, and thereby contributing to vibration suppression in the structures [34]. However, Newtonian fluids show a loss of performance under severe vibrational loadings, as Newtonian fluids do not show thicker textures at higher shear rates, and thereby leading to spill out and leakage from the structures. However, STF compensates the inefficiency of the Newtonian fluids even at severe conditions. Due to its adaptive rheology, STF is a promising smart material for self-activated passive vibration isolation structures.

5.5 Conclusions

Vibration damping is a challenging issue in engineering structures. There are several attempts for controlling vibrational loadings in engineering structures. Smart applications are one of these attempts that benefit from advanced systems in vibration damping. In recent years, STF has emerged as a promising material for smart vibration damping applications. Non-Newtonian rheology of STF provides adaptive damping properties for smart applications.

STF has been used in various forms in vibration damping systems, such as bulk liquid in containers, impregnating agent in porous structures, coating material on layered assemblies, etc. This provides ease of use to adapt STF to different types of structures. The adoption of STF-based vibration damping solutions promises to redefine the landscape of engineering applications. By providing effective and adaptable means to mitigate vibrations, STFs offer the potential to enhance passenger comfort, improve structural integrity, elevate athletic performance, and extend the operational lifespan of machinery. The widespread integration of STFs into various industries heralds a new era of engineering innovation.

References

1. Zuo, H., Bi, K., Hao, H.: A state-of-the-art review on the vibration mitigation of wind turbines. Renew. Sust. Energ. Rev. **121**, 109710 (2020)
2. Zhou, X.Q., Yu, D.Y., Shao, X.Y., Zhang, S.Q., Wang, S.: Research and applications of viscoelastic vibration damping materials: a review. Compos. Struct. **136**, 460–480 (2016)
3. Wagner, N.J., Brady, J.F.: Shear thickening in colloidal dispersions. Phys. Today. **62**(10), 27–32 (2009)
4. Gürgen, S., Li, W., Kuşhan, M.C.: The rheology of shear thickening fluids with various ceramic particle additives. Mater. Des. **104**, 312–319 (2016)
5. Gürgen, S., Sofuoğlu, M.A.: Experimental investigation on vibration characteristics of shear thickening fluid filled CFRP tubes. Compos. Struct. **226**, 111236 (2019)
6. Wei, M., Lin, K., Sun, L.: Shear thickening fluids and their applications. Mater. Des. **216**, 110570 (2022)
7. Galindo-Rosales, F.: Complex fluids in energy dissipating systems. Appl. Sci. **6**(8), 206 (2016)
8. Mari, R., Seto, R., Morris, J.F., Denn, M.M.: Shear thickening, frictionless and frictional rheologies in non-Brownian suspensions. J. Rheol. **58**(6), 1693–1724 (2014)
9. Gürgen, S., Kuşhan, M.C., Li, W.: Shear thickening fluids in protective applications: a review. Prog. Polym. Sci. **75**, 48–72 (2017)
10. Liu, D.M.: Particle packing and rheological property of highly-concentrated ceramic suspensions: φm determination and viscosity prediction. J. Mater. Sci. **35**(21), 5503–5507 (2000)
11. Srivastava, A., Majumdar, A., Butola, B.S.: Improving the impact resistance of textile structures by using shear thickening fluids: a review. Crit. Rev. Solid State Mater. Sci. **37**(2), 115–129 (2012)
12. Gürgen, S., Sofuoğlu, M.A.: Smart polymer integrated cork composites for enhanced vibration damping properties. Compos. Struct. **258**, 113200 (2021)
13. Fu, K., Wang, H., Zhang, Y.X., Ye, L., Escobedo, J.P., Hazell, P.J., et al.: Rheological and energy absorption characteristics of a concentrated shear thickening fluid at various temperatures. Int. J. Impact Eng. **139**, 103525 (2020)
14. Gürgen, S., Sofuoğlu, M.A., Kuşhan, M.C.: Rheological compatibility of multi-phase shear thickening fluid with a phenomenological model. Smart Mater. Struct. **28**(3), 035027 (2019)
15. Zhou, H., Yan, L., Jiang, W., Xuan, S., Gong, X.: Shear thickening fluid–based energy-free damper: design and dynamic characteristics. J. Intell. Mater. Syst. Struct. **27**(2), 208–220 (2016)
16. Sheikhi, M.R., Gürgen, S., Altuntas, O., Sofuoğlu, M.A.: Anti-impact and vibration-damping design of cork-based sandwich structures for low-speed aerial vehicles. Arch. Civ. Mech. Eng. **23**(2), 71 (2023)
17. Lin, K., Zhou, A., Liu, H., Liu, Y., Huang, C.: Shear thickening fluid damper and its application to vibration mitigation of stay cable. Structure. **26**, 214–223 (2020)
18. Wei, M., Hu, G., Jin, L., Lin, K., Zou, D.: Forced vibration of a shear thickening fluid sandwich beam. Smart Mater. Struct. **25**(5), 055041 (2016)
19. Yeh, S.K., Lin, J.J., Zhuang, H.Y., Chen, Y.C., Chang, H.C., Zheng, J.Y., et al.: Light shear thickening fluid (STF)/Kevlar composites with improved ballistic impact strength. J. Polym. Res. **26**(6), 155 (2019)

20. Sun, L., Wang, G., Zhang, C., Jin, Q., Song, Y.: On the rheological properties of multi-walled carbon nano-polyvinylpyrrolidone/silicon-based shear thickening fluid. Nanotechnol. Rev. **10**(1), 1339–1348 (2021)
21. Huang, W., Wu, Y., Qiu, L., Dong, C., Ding, J., Li, D.: Tuning rheological performance of silica concentrated shear thickening fluid by using graphene oxide. Adv. Condens. Matter. Phys. **2015**, 1–5 (2015)
22. Gürgen, S., Kuşhan, M.C., Li, W.: The effect of carbide particle additives on rheology of shear thickening fluids. Korea-Aust. Rheol. J. **28**(2), 121–128 (2016)
23. Galindo-Rosales, F.J., Martínez-Aranda, S., Campo-Deaño, L.: CorkSTFμfluidics – a novel concept for the development of eco-friendly light-weight energy absorbing composites. Mater. Des. **82**, 326–334 (2015)
24. Gürgen, S., Fernandes, F.A.O., de Sousa, R.J.A., Kuşhan, M.C.: Development of eco-friendly shock-absorbing Cork composites enhanced by a non-Newtonian fluid. Appl. Compos. Mater. **28**(1), 165–179 (2021)
25. Gürgen, S., De Sousa, R.J.A.: Rheological and deformation behavior of natural smart suspensions exhibiting shear thickening properties. Arch. Civ. Mech. Eng. **20**(4), 110 (2020)
26. Ghosh, A., Majumdar, A., Butola, B.S.: Role of surface chemistry of fibres additives on rheological behavior of ceramic particle based shear thickening fluids. Ceram. Int. **44**(17), 21514–21524 (2018)
27. Gürgen, S., Sofuoğlu, M.A.: Vibration attenuation of sandwich structures filled with shear thickening fluids. Compos. Part B Eng. **186**, 107831 (2020)
28. Nakonieczna, P., Wierzbicki, Ł., Wróblewski, R., Płociński, T., Leonowicz, M.: The influence of carbon nanotube addition on the properties of shear thickening fluid. Bull. Mater. Sci. **42**(4), 162 (2019)
29. Gong, X., Zhang, J., Xuan, S.: Multi-functional systems based on shear thickening fluid. In: Gürgen, S. (ed.) Shear Thickening Fluid [Internet], pp. 53–75. Springer, Cham (2023) Available from: https://link.springer.com/10.1007/978-3-031-25717-9_4 [cited 2023 Aug 16]
30. Lin, N.Y.C., Guy, B.M., Hermes, M., Ness, C., Sun, J., Poon, W.C.K., et al.: Hydrodynamic and contact contributions to continuous shear thickening in colloidal suspensions. Phys. Rev. Lett. **115**(22), 228304 (2015)
31. Hasanzadeh, M., Mottaghitalab, V.: Tuning of the rheological properties of concentrated silica suspensions using carbon nanotubes. Rheol. Acta. **55**(9), 759–766 (2016)
32. Decker, M.J., Halbach, C.J., Nam, C.H., Wagner, N.J., Wetzel, E.D.: Stab resistance of shear thickening fluid (STF)-treated fabrics. Compos. Sci. Technol. **67**(3–4), 565–578 (2007)
33. Maranzano, B.J., Wagner, N.J.: The effects of interparticle interactions and particle size on reversible shear thickening: hard-sphere colloidal dispersions. J. Rheol. **45**(5), 1205–1222 (2001)
34. Shum, K.M.: Closed form optimal solution of a tuned liquid column damper for suppressing harmonic vibration of structures. Eng. Struct. **31**(1), 84–92 (2009)

Chapter 6
Smart Polishing with Shear Thickening Fluid

Boyang Wan, Ziyan Man, and Li Chang

6.1 Introduction

Manufacturing processing holds a central position in the mechanical engineering, substantially impacting the quality, functionality, and overall performance of all engineering products. It is the basic production method to transfer raw material into the finished product with the desired geometry shape and to render the material's internal structures/properties more suitable for specific applications. By employing a variety of machining processing techniques, manufacturers can achieve desired surface characteristics, including better durability, enhanced aesthetics, superior resistance to corrosion, optimized friction and wear characteristics, and improved functionalities.

Traditional manufacturing processing techniques encompass a range of methods such as grinding, polishing, shot peening, turning, milling, lapping, superfinishing, and honing. However, these conventional processes present challenges in terms of process control and sometimes the generation of deep cracks due to the application of high contact forces. In this case, achieving consistent and deterministic machining outcomes becomes difficult. In order to enhance the quality of surface finishing, the utilization of the shear thickening fluid (STF) has emerged as a promising approach in recent years [1]. The STF is a non-Newtonian fluid typically composed of micro or nanoparticles dispersed in a suspension. The STF exhibits a liquid state under low shear rates, while it transforms into a highly viscous and sticky substance, even a solid state under high shear rates. Galindo-Rosales et al. [2] illustrated the viscosity of STF at various shear rates in Fig. 6.1.

B. Wan · Z. Man · L. Chang (✉)
School of Aerospace, Mechanical and Mechatronic Engineering, The University of Sydney, Sydney, NSW, Australia
e-mail: li.chang@sydney.edu.au

S. Gürgen (ed.), *Smart Systems with Shear Thickening Fluid*,
https://doi.org/10.1007/978-3-031-53570-3_6

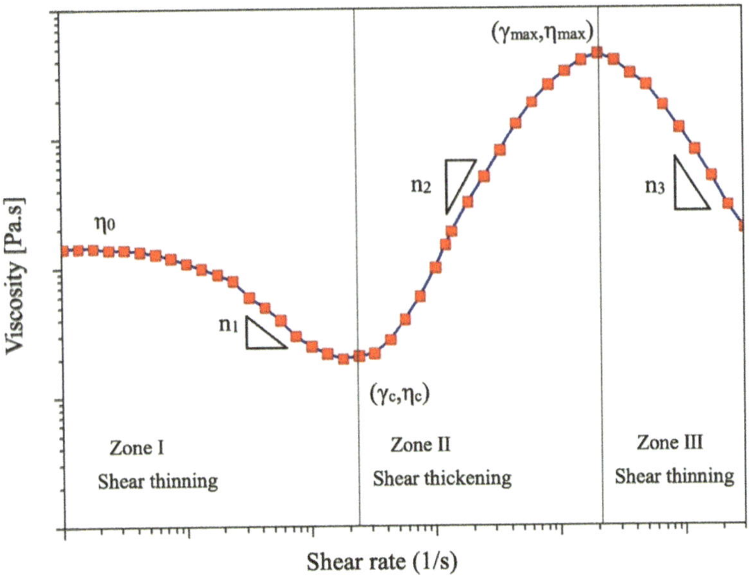

Fig. 6.1 Typical viscosity curve of STF over the change of shear rate [3]

The shear thickening process can be divided into three different stages: shear thinning (Zone I), followed by a rapid increase in viscosity (Zone II), and concluding with another stage of shear thinning (Zone III) at high shear rates. To unravel its rheological behavior and underlying mechanics, Cwalina and Wagner [4] proposed a theory explaining shear thickening based on the formation of particle clusters. When the shear rate exceeds a critical value, the formation of particle clusters leads to the formation of macroscopic solid-like features in STF. This has inspired new, smart manufacturing methods by using the STF as a self-adaptive material, such as shear thickening polishing (STP) [1]. Given the distinctive characteristics of STFs, researchers and engineers have embarked on exploring novel approaches to enhance the efficiency, precision, and effectiveness of the surface finish [5–8]. The objective of this chapter is to conduct a state-of-art literature review, aiming to deepen our understanding of the mechanisms in the STP process, and to explore other applications relevant to material processing/treatment within the realm of smart manufacturing. The findings of this chapter could be used to improve the current processing systems and expand the utilization of STFs for smart manufacturing, thereby contributing to advancements in the field.

6.2 Applications of STF in Manufacturing Processes

In recent years, there have been significant efforts to comprehend the rheological behavior of STFs and explore their applications in various manufacturing processes. This session aims to provide an overview of recent progresses in utilizing STFs for three manufacturing application streams: (i) surface polishing, (ii) machining process optimization, and (iii) surface/structure modification purposes.

6.2.1 Shear Thickening Polishing

STP is an emerging surface finishing technique which draws significant attention in recent years [1, 5–8]. Compared to the conventional mechanical polishing methods, STP is less dependent on the geometric features of the workpiece and does not require a polishing pad. By combining the STF with abrasive particles or slurries, a shear thickening abrasive media can be formulated, also named the "flexible fixed abrasive tool." This media proves instrumental in achieving exceptional polishing results on complex structures, even with curved surfaces.

6.2.1.1 The Fundamental Principle of STP

STP is an innovative non-contact polishing method that harnesses the shear thickening effect of non-Newtonian slurries [5]. During the STP process, when the workpiece is submerged in the shear thickening slurry, the flow velocity gradient near the workpiece triggers the shear thickening effect. The fundamental principle of the STP process is illustrated schematically in Fig. 6.2 [6].

The STP slurry consists of a dispersed media, solid colloidal particles, and abrasives. The workpiece and the STP slurry experience relative motion, leading to the shear thickening phenomena and resulting in a sudden increase in shear stress near the contact area. Subsequently, the abrasive particles efficiently remove the materials from the workpiece surface.

During the initial stage, the solid particles and abrasives are uniformly dispersed within the liquid matrix in the STP slurry. Under relatively low shear rates, the motion of abrasives in the slurry resembles that of a Newtonian fluid. Due to insufficient dynamic force acting on the abrasives, most of them simply rotate or flip over the roughness peaks on the workpiece without effectively cutting. As a result, no material removal from the workpiece occurs during this stage.

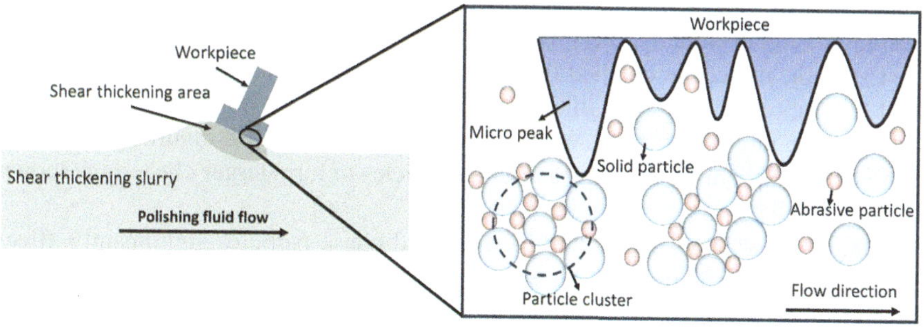

Fig. 6.2 Mechanism of STP process [7]

As the shear rate increases, shear thickening is triggered, causing the STP slurry to transition into a solid phase. As described in the previous session, solid particles in the slurry rapidly aggregate, forming large particle clusters that encapsulate the abrasive particles. Consequently, a significant shear force is generated, enabling the abrasive particles to cut through the roughness peaks instead of rolling over them. When the shear force surpasses the resistance force provided by the roughness peak, the upper part of the peak is sheared off from the workpiece. With a further increase in the relative velocity of fluid flow, the dynamic force acting on the abrasive particles intensifies, resulting in enhanced shear thickening and subsequent removal of more peaks. Once the target peak has been sheared off, the reduced surface roughness leads to a decrease in shear stress and a milder shear thickening effect. Eventually, the particle clusters disintegrate as a smooth surface area is achieved. Upon the removal of the shearing force, the slurry reverts to its original flow properties. The final polishing performance is commonly evaluated according to the surface topography, e.g., by using confocal microscopy techniques.

6.2.1.2 STF Materials Used for STP Applications

The preparation of a high-performance slurry with an optimal shear thickening effect is fundamental in the applications of STP. Among various preparation systems, the granular suspension system is commonly employed for polishing purposes [9]. Granular suspensions with high concentrations often exhibit a pronounced shear thickening effect, resulting in a more notable transformation of the fluid into a solid-like state compared to other preparation systems, such as cross-linked polymer and micelle systems. The granular suspension system in STFs comprises several key components, namely the dispersant, dispersed phase particles, and, optionally, abrasive particles, which are summarized in Table 6.1. This section aims to provide an overview of these components within the granular suspension system.

The molecular weight, viscosity, and steric hindrance of the suspension media have certain effects on the rheological properties of the STF. Many dispersant fluids have been investigated, including deionized (DI) water, polyethylene glycol (PEG), and ethylene glycol (EG). Among them, DI water and PEG are the most widely used dispersant of STP slurry owing to the advantages of polarity and stability. Jiang et al. [26] used a mixture of glycerol and water as the dispersion medium, and they found that by increasing the ratio of glycerol and water, the critical shear rate decreased, and eventually, the shear thickening effect was enhanced. Furthermore, the rise in the molecular chain of the dispersion media can increase the viscosity and lead to a more evident shear thickening effect. As a result, compared to EG, PEG has a longer molecular chain, which leads to a stronger steric hindrance, and it is more likely to promote the aggregation of particles to form larger clustering (shown in Fig. 6.2).

Besides the dispersion media, the dispersed phase particles significantly affect the rheological properties of the STF. The first dispersed phase of STP slurry is generally selected from solid colloidal particles, including natural minerals, metals,

Table 6.1 Common materials used in the preparation of STP slurry

	Materials	References
Dispersed media	Water	[6, 10–15]
	Ethylene glycol (EG)	[16]
	Polyethylene glycol (PEG)	[4, 17–20]
First dispersed phase	Fumed silica	[4, 8, 12, 18]
	Multi-hydroxyl polymers	[11–15, 21–24]
	Iron, Fe	[25]
Abrasive dispersed phase	Boron carbide, B_4C	[17]
	Aluminum oxide, Al_2O_3	[10–13, 17, 24]
	Silicon carbide, SiC	[8, 17, 19, 20, 25]
	Silicon oxide, SiO_2	[5, 15, 20]
	Diamond	[14, 23]

and polymers, with particle size within the range of 1 nm to 1 μm. Among them, corn starch is widely employed by a number of researchers to explore the mechanism of shear thickening and jamming [27, 28]. In addition, synthetic polymer particles exhibit a unique advantage owing to the controllable particle size and shape (their effects will be discussed in the following section) [27]. Despite the success of the polymer-based dispersed phase, due to the special requirements in abrasive bearing capacity for the polishing purpose, mineral and metal particles are more commonly used in the preparation of STFs.

To further reinforce the polishing quality and efficiency of STFs, the fine abrasive particles were added to the preconfigured STFs as a second dispersed phase. The addition of abrasive particles increases the volume fraction of dispersed particles in the original STFs. Then it alters the formation of particle clusters during the polishing. For instance, Gürgen et al. [17] found that the ceramic nanoparticles could shorten the time of cluster formation. Sun et al. [29] observed a larger clustering by adding Nd_2O_3 nanoparticles to the SiO_2-based STFs.

6.2.2 *Applications of STFs in Smart Manufacturing Operations*

The turning machining method is widely employed in workpiece machining, encompassing cutting, and milling processes that involve collisions and interactions between the workpiece and the tool. The elasticity of both the workpiece and tools can cause changes in their relative positions, resulting in vibrations that can lead to machining errors and undesirable surface finishes. Surface quality is paramount in manufacturing processes as it directly impacts the functionality and performance of structures. In this regard, many researchers attempted to mitigate the vibration during the machining operations by various methods. STFs have been recently used to suppress chatter instabilities and forced vibrations in machining operations. It

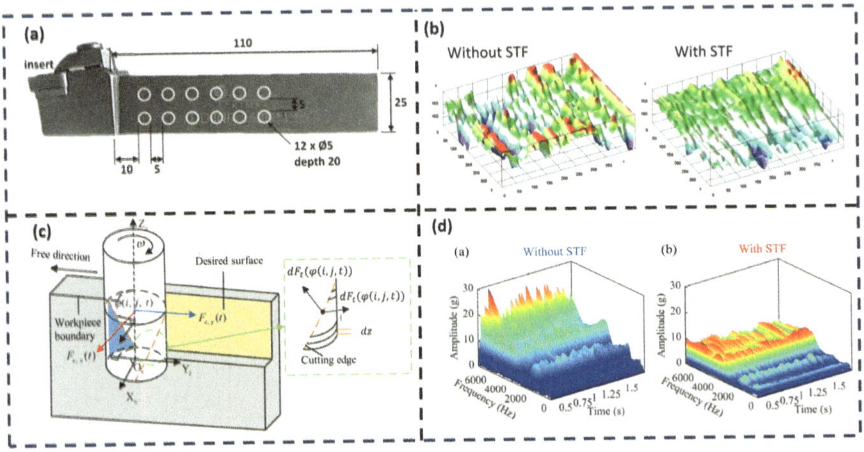

Fig. 6.3 (**a**) STF reinforced cutting tool [18], (**b**) Surface topographies of the workpiece with different cutting tools [18], (**c**) Milling process in the thin-walled workpiece [30], and (**d**) the vibration amplitude of the thin-walled workpiece with and without STF in the milling process [30]

should be noted that such an application of STFs in machine processing is a relatively new field. Therefore, there is still a gap in taking full advantage of STFs in damping control during manufacturing operations.

An early attempt was made by Gürgen and his colleagues in 2020 [18]. They drilled a series of holes in the cutting tool and filled them with an STF, as shown in Fig. 6.3a. Modal analysis was employed to determine the damping ratio of the cutting tools, and the results indicated that the addition of the STF improved the damping capacity of the cutting tool. Furthermore, Gürgen et al. [18] conducted a set of machining tests and measured the surface roughness of the workpieces to gain a comprehensive understanding of the role of the STF in cutting tools. According to the measurements (Fig. 6.3b), the integration of the STF resulted in a smoother surface finish, with surface roughness, R_a, of 1.12 μm, compared to the conventional tool (without STF), R_a, of 1.27 μm. Consequently, it can be concluded that incorporating STFs into cutting tools can reduce vibrations and damping during operations, thereby enhancing surface accuracy and machining performance.

Another notable example comes from Yuan et al. [30], who proposed an STF-based system for milling the thin-walled structures, as shown in Fig. 6.3c. Thin-walled structures are prone to forced vibrations due to their low stiffness. To address this issue, Yuan et al. [30] utilized the STF filler to support the thin wall structures. Their results demonstrated that vibration amplitudes were significantly suppressed when using STF support during milling operations (Fig. 6.3d). This led to a more stable surface finish and chatter-free operations. It is important to note that the vibrational modes of thin-walled workpieces can be highly complex, influenced by factors such as cutter position, cutting force frequency, and magnitude. Nevertheless, STF-based systems provide a practical approach for chatter suppression and exhibit promising prospects in smart manufacturing applications.

6.2.3 Fabricating Smart Surfaces Treated with STFs

Extensive research has been conducted on incorporating STFs into the matrix of composites to enhance their mechanical properties and energy absorption behavior. STF-reinforced composites offer superior protection against impact and scratch loadings, making them suitable for various applications such as ballistic resistance. As demonstrated in Fig. 6.4a, various treatment methods have been explored for STF-reinforced composites, including impregnation, dipping, and coating. Impregnation and filling methods involve immersing STF into the fibers and fabric throughout the entire structure to achieve optimal enhancement performance. However, these methods often result in a significant increase in the weight of the composite. On the other hand, STF-coating formulations are applied only to the surface of materials, allowing for surface modification without a substantial weight increase.

Kim et al. [32] demonstrated that a surface treated with STF using a spraying-based method could increase energy absorption by 90% compared to untreated surfaces, with less than a 15% increase in weight, as shown in Fig. 6.4b. This study highlights the feasibility and effectiveness of STF surface treatment for bulletproof applications. However, some researchers have expressed concerns about the stability of STF coatings due to poor permeability, which can result in the coating easily coming off [31]. In addition to spraying, surface treatments using atomic plasma or chemical deposition can enhance the stability of STF coatings by increasing the level of friction.

Furthermore, the combination of nanoparticles, magnetic particles, conductive particles, and other additives with STF has shown potential for creating smart surfaces with shape-adapting and self-healing properties [33–35]. Nevertheless, these methods often involve heating, etching, and chemical treatments, which may potentially degrade the overall mechanical properties of the fiber and matrix. Therefore, it is necessary to further investigate the working mechanisms and overall performance of STF surface treatment methods.

6.3 Design of STFs for Manufacturing Applications

In general, the applications of the STF in smart manufacturing rely on its unique rheological behavior, i.e., the STF can display unusual phase transitions between liquid and "solid" phases depending on the operation conditions. In practice, the viscosity and shear behavior of STFs can be first characterized by the rheometer in the lab. Accordingly, the working parameters can be further selected to control the desired responses of the STF under different working conditions, e.g., the STP slurry needs to operate under the shear thickening conditions (Zone II in Fig. 6.1). According to the available literature, many factors could affect the performance of STFs, such as STF viscosity, particle concentration, operation velocity, boundary

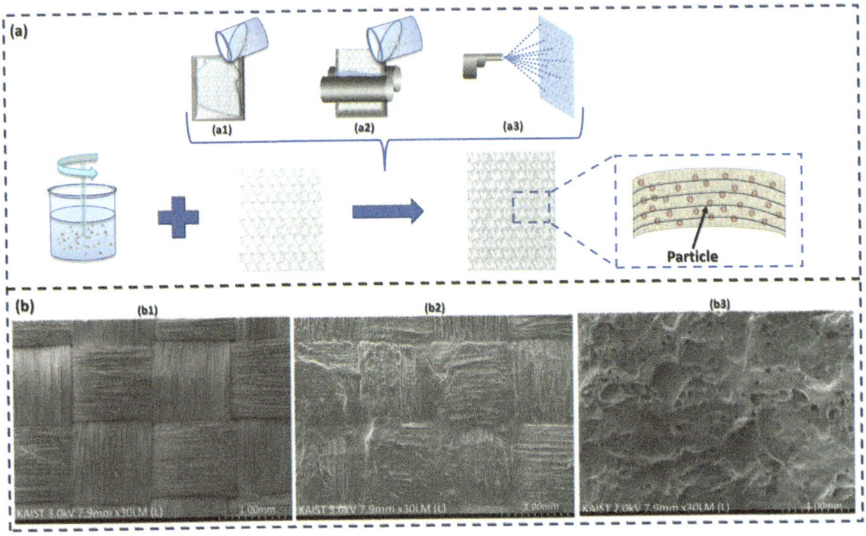

Fig. 6.4 (**a**) Fabrication methods for STF-reinforced composite: (**a**1) impregnation, (**a**2) dipping extrusion, and (**a**3) spray gun [31], (**b**) SEM images of the fabrics: (**b**1) without STF treatment, (**b**2) STF impregnated, and (**b**3) STF coated [32]

conditions, etc. In this session, we will focus on how to design an effective STP system by selecting two groups of parameters, i.e., material parameters of STF slurry and manufacturing operation parameters.

6.3.1 Development of STF Slurries for STP

According to Gürgen et al.'s work, as shown in Fig. 6.5a, by comparing the 2D surface view and corresponding 3D topography of the specimen before and after the STF polishing, it can be seen that the group of "no thickening" exhibits a similar average surface roughness at the beginning and end of the process, which indicates there is no effective material removal from the workpiece [8]. Therefore, Gürgen et al. suggested that it is essential to make sure the STF slurry is in the shear thickening region to ensure the polishing outcomes. As explained in the previous section, to increase the polishing efficiency, the abrasive particles are added to the STF slurry. Figure 6.5b shows a typical SEM image for the abrasive powders. Gürgen et al. [8] proved that the existence of abrasive particles enhances the viscosity of the STF slurry under the given shear rate (Fig. 6.5b), and the viscosity of STF slurry increases with the abrasive concentrations (Fig. 6.5c), which means that the shear thickening performance can be effectively improved by increasing abrasives

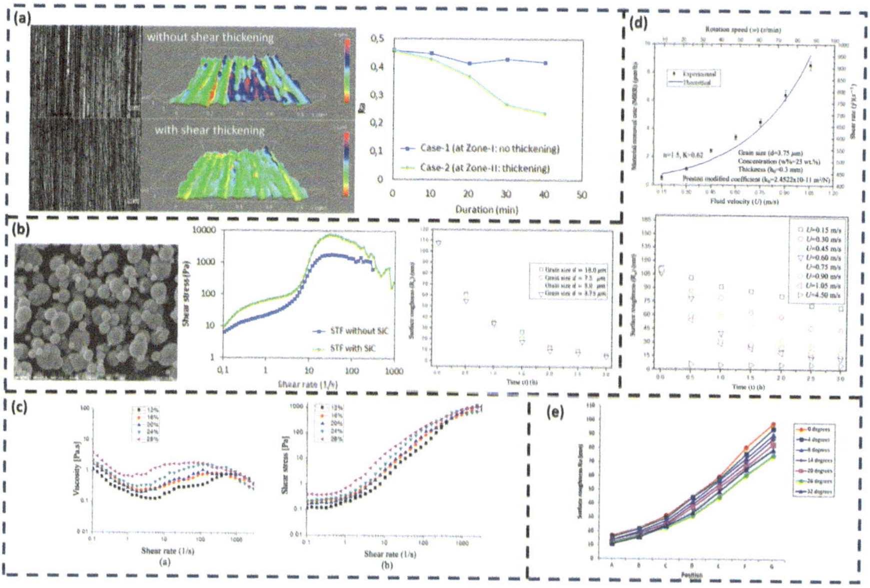

Fig. 6.5 (a) the polishing performance of STF [8], and (b)–(e) main influencing factors of STP: (b) grain size [3, 6, 8], (c) abrasive particle concentrations [3], (d) polishing velocity [6] and (e) inclination angle [14]

concentration [3]. However, it is worth noting that when the abrasive concentration exceeds some certain limit, the polishing outcome could be impaired due to the uneven agglomeration of particles which could damage the surface.

In addition to the concentration of the abrasive, some researchers argue that the shape and size of additives are another contributing factor [9, 36]. According to the study by Chang et al. [37], the finer abrasive particles could result in a smoother surface. This observation is consistent with the work by Lyu et al. [23]. However, according to the work from Li. et al. [6], the grain size only has a marginal effect on the average surface roughness of polished surfaces. It could be explained that the resultant cutting force and normal pressure primarily depend on the macroscopic mechanical properties of the suspension. Therefore, this may need further investigation to determine the effect of grain size as well as the morphology of solid colloidal particles on STP process.

6.3.2 Selection and Control of the Operational Parameters

Besides the inherent rheological behaviors of STF slurries, operational parameters are another crucial contributing factors that affect polishing outcomes, such as polishing speed, inclination angle, and working distance. For example, Li et al. [6] examined the influence of polishing velocity on the material removal rate and surface roughness. Both the experimental and analytical results showed that there was a power-function relationship between the material removal rate and the polishing velocity (Fig. 6.5d). As a result, the surface roughness of the workpiece had a gradual decrease with processing time and finally achieved a finer surface.

The inclination angle is another often discussed contributing factor, particularly in polishing complex-shaped workpieces [14, 38, 39]. Nguyen and his colleagues [14, 24, 39] conducted a series of works to optimize the inclination angle for polishing several geometries, including steel gear and hip implant heads. Their results suggested that the optimum inclination angle is dependent on the workpiece geometries. Generally, the surface roughness decreases with the rise of the angle. However, the excessive inclination angle could lead to an uneven surface profile indicating a poor finishing quality (Fig. 6.5e). Apart from polishing velocity and angle, the distance between the workpiece and the inner boundary of the polishing tank is also a factor that affects the surface roughness because it influences the pressure distribution on the workpiece surface.

In conclusion, the optimum operational parameters can enhance the shear thickening effect, thereby enabling more effective abrasives and strengthening the cutting ability of the abrasive clusters. Thus, surface quality improvement would be attributed to more micro-peaks being removed from the workpiece with the same processing time. It can be concluded from various studies that the polishing speed, inclination angle, and work distance are the main machine processing parameters, which have the most significant effect on the surface roughness of the workpiece.

6.4 Recent Advances in Modeling of STFs for Manufacturing Applications

Owing to the complex transfer behavior of STFs, analytical approaches to describe the flow behavior and rheological properties have been proven difficult. In response, researchers have turned to developing empirical models based on experimental data to predict the behavior of STFs. These empirical models aim to capture the intricate relationship between various parameters, such as shear rate and stress, as well as their impact on the shear thickening phenomenon. According to the proposed empirical models, a number of researchers utilized computational tools to investigate the applications of STFs in the manufacturing field. Notably, computational modeling of the STP process has garnered significant attention in recent years. This section will delve into the latest advancements and endeavors made in the modeling of STFs for their application in manufacturing.

6.4.1 Empirical Modeling

Numerous researchers have dedicated considerable efforts to investigating the material removal mechanism in STP [6, 40, 41]. In this regard, the application of the Preston equation, initially utilized as an empirical formula in the grinding process, has been extended to the STP process. The material removal rate (MRR) can be expressed in Eq. (6.1) [42]:

$$MRR = k \cdot p \cdot u_0 = k \cdot p \cdot 2\pi\omega l \tag{6.1}$$

where $k = k_1 \cdot k_2 \cdot k_3$ is modified Preston coefficient; k_1, k_2, and k_3 are the coefficient related to the workpiece material properties and the coefficient related to STF properties, respectively. p is pressure on the workpiece surface, and u_0 is the relative velocity of fluid flow, which is associated with the plate rational speed, ω, and the distance between the boundary and the polishing surface of the workpiece, l.

The pressure, p, of STF on a workpiece is a complex parameter. To simplify its expression, it can be approximately estimated as the fluid pressure, p_d, shown as Eq. (6.2) [42, 43],

$$p_d = \alpha \cdot U^n \cdot f(n, \varphi) \tag{6.2}$$

Where U is fluid velocity provided by the polishing plate, α and n are constant coefficients. $f(n, \varphi)$ is a function of n and φ, which represent the geometrical parameters of workpieces.

The determination of the Preston-modified coefficient is important to elucidate the material removal model in STP. To achieve this, rheological tests are conducted to determine the material properties once the STP slurry is prepared. The calculation of the Preston coefficient is then carried out based on the specific workpiece geometry and operating conditions, such as rotation speed and fluid velocity.

6.4.2 Computational Simulation

Computational models in this field generally fall into two main categories: continuum-based approaches and particle-based approaches (Fig. 6.6a). The continuum-based approach, commonly employed in computational fluid dynamics (CFD) analysis, is a well-established method for modeling the flow of STFs [44]. As discussed above, viscosity has become a key index for realizing STF rheological-mechanical behaviors. The power-law intrinsic equation is the most commonly used function to fit the viscosity data of STFs, and it is represented as Eq. (6.3):

$$\mu(\dot{\gamma}) = k\dot{\gamma}^{n-1} \tag{6.3}$$

where μ is the viscosity of STF, which is a function of shear rate, $\dot{\gamma}$. k is the power law coefficient, and n is the fluidity index.

A number of researchers conducted CFD simulations to investigate the relationship between the machining parameters, such as the working distance, velocity and inclination angle for the gear surface, and the polishing response (surface pressure and shear stress) based on a simplified 2D model [11, 14].

As the demand for considering more complex workpiece surface geometries has increased, commercial software packages like ANSYS and COMSOL have enabled the coupling of CFD solvers with finite element analysis (FEA) frameworks (Fig. 6.6b). Shao et al. [15] and Nguyen et al. [39] investigated STF flow with complex 3D surfaces by using ANSYS Fluent module. The simulation results by Nguyen et al. [39] suggest that the surface pressure is significantly influenced by the inclination angle. The optimal inclination angle range of 26–45 degrees was found to achieve the highest and most uniformly distributed pressure on the workpiece surface. Additionally, other researchers [20, 22, 25] utilized COMSOL software to consider the effects of various physical systems, such as temperature and magnetization, on the STF owing to the advantages of performing the multi-phasic simulation.

Despite the advancements made with the CFD-FEA framework, there are still debates regarding the unique shear thickening behavior of colloidal suspensions due to the presence of multiple phases. For instance, the influence of micro-particles on the macroscopic shear thickening phenomenon in suspensions remains unclear. Some researchers have observed discontinuous shear thickening when the volume

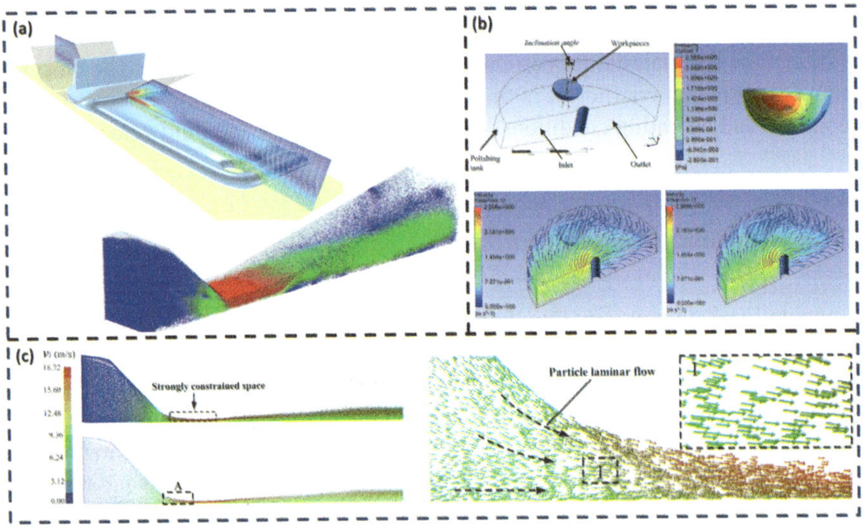

Fig. 6.6 (**a**) Computational simulation of STF by using continuum CFD and discrete particle method, (**b**) a typical FE model of STF polishing process [24], and (**c**) characteristics of particle motion state by using DEM [10]

fraction of abrasive particles exceeds a critical value, instead of a continuous viscosity increase at any applied shear rate [45]. To address these research questions, particle-based approaches such as the discrete element method (DEM) and molecular dynamics (MD) simulations have been introduced into the field of STP.

Ji et al. [46] developed a discrete phase model to investigate the dynamic suspension model of two-phase abrasive flow. Simulation results indicate that the abrasive flow polishing process mainly exhibits translation of ablating location, which is affected by the granular pressure and near-wall velocity. Recent research, with the help of advanced computational tools, found that the discontinuous shear thickening observed is attributed to fractional contact between dispersed particles, which is an independent phenomenon distinct from shear-jammed traditions [47].

Mai-Duy et al. [48] utilized the particle dynamic method to explore the impact of the size and volume fraction of dispersed particles on the shear thickening behavior during polishing. Kroupa et al. [49] treated abrasive particles as Hertz model and investigated the particle interaction with flow based on the lubrication theory by using DEM. Similarly, Ge et al. [10] calculated the particle-wall collision on the workpiece surface based on the DEM-CFD coupled method (Fig. 6.6c). Chen et al. [50] employed MD simulation to study the fundamental shear thickening mechanism by focusing on the evolution of particle clusters. In addition, smooth particle hydrodynamic (SPH) method is another widely used computational tool to study the material removal behavior during material processing [51, 52]. This method is particularly suitable for modeling complex fluid-solid interaction due to its unique advantages of handling large deformation and free surface flows.

Numerous efforts have been made to develop an accurate computational model for describing the viscosity function of multi-phase suspensions. However, the proposed model is exceedingly complex and needs a significant computational expense. Owing to the unique advantage of machine learning, such as being highly effective and efficient for modeling and optimizing complex relationships, it has been recently introduced to modeling of multi-phase STF suspension. Some researchers proposed nonparametric artificial neural network models, which determined the relationship between input and output parameters by training the experimental dataset with the mapping technique [19]. Although the fitting results of these models to experimental data may not align well under certain conditions, potentially due to insufficient training data, they present a promising method for the development of STF calculation models.

6.5 Summary, Challenges, and Perspectives

This chapter presents a comprehensive review of recent advancements in smart manufacturing applications utilizing STFs. The applications of STFs in manufacturing, such as surface polishing, have demonstrated unique advantages over conventional machining processes. For instance, numerous experimental studies have showcased the effectiveness of polishing complex structures with curved surfaces, which cannot be easily achieved by conventional polishing methods. In this context, the main influencing factors on the performance of STFs for those applications are further summarized and discussed in detail. In particular, theoretical modeling has played a vital role in deepening our understanding of the underlying mechanisms, and facilitating further advancements in the field. Based on the theoretical model, some numerical studies are carried out to investigate stress distribution, rheological flow, and material removal behavior. Finally, some current challenges and future perspectives associated with STFs in surface processing are summarized as follows:

1. Owing to the development of more advanced computational techniques, it becomes feasible to analyze the material removal behavior at a finer level. However, it is still challenging to precisely model the material removal process and fully understand the machining mechanism with the STFs. Recently, shear thickening effect has been divided into two categories: continuous shear thickening and discontinuous shear thickening, each associated with distinct mechanisms. Continuous shear thickening is commonly attributed to the "clustering" mechanism, whereas discontinuous shear thickening is explained by the "friction contact rheology" model. A multi-scale modeling framework shows promising potential to thoroughly investigate the relationship between the microscopic particle interaction and jamming and the macroscopic rheological behaviors. Understanding complex physical phenomena, such as fluid-solid phase transfer, during the polishing process is necessary for developing efficient and effective surface processing strategies.
2. The intrinsic rheological properties of STFs impose constraints on the shear stress during polishing and other machining procedures, subsequently impacting the efficiency and quality of surface finishes in industrial applications. These inherent limitations have hindered the widespread adoption of STP in the market. As a result, research into STP remains primarily in the developmental and investigational phase at present. To improve the above-mentioned limitations, two or more compliant machining processes can be potentially hybridized, with the expectation of enhancing machining accuracy, quality, stability, and productivity. One successful implementation is using magnetic fields to enhance the shear thickening field and therefore improve the polishing performance. Although the precise mechanisms behind shear thickening under the magnetic field are not yet fully understood, magnetorheological STFs has demonstrated great potential in the field of ultra-precision machining, particularly in complex surface polishing processes.
3. By incorporating the concept of smart manufacturing, adaptive STP technologies are highly demanded for the next decade. One promising direction for future development could be the development of in-process sensors, which is beneficial for monitoring the polishing process, thereby optimizing the operational parameters to achieve an ultra-precision surface finish of workpieces with relatively sophisticated geometry, such as complex curvature and thin-walled structure.

References

1. Chang, L., Friedrich, K., Ye, L.: Method, systems and compositions for polishing. International Patent WO/2013/016779 (2013)
2. Galindo-Rosales, F., et al.: How Dr. Malcom M. Cross may have tackled the development of "An apparent viscosity function for shear thickening fluids". J. Non-Newtonian Fluid Mech. **166**(23–24), 1421–1424 (2011)
3. Ren, Y., et al.: Research on the rheological characteristic of magnetorheological shear thickening fluid for polishing process. Int. J. Adv. Manuf. Technol. **117**, 413–423 (2021)
4. Cwalina, C.D., Wagner, N.J.: Rheology of non-Brownian particles suspended in concentrated colloidal dispersions at low particle Reynolds number. J. Rheol. **60**(1), 47–59 (2016)
5. Li, M., et al.: Surface integrity of bearing steel element with a new high-efficiency shear thickening polishing technique. Procedia CIRP. **71**, 313–316 (2018)
6. Li, M., et al.: Shear-thickening polishing method. Int. J. Mach. Tools Manuf. **94**, 88–99 (2015)
7. Shao, Q., et al.: Shear thickening polishing of quartz glass. Micromachines. **12**(8), 956 (2021)
8. Gürgen, S., Sert, A.: Polishing operation of a steel bar in a shear thickening fluid medium. Compos. Part B. **175**, 107127 (2019)
9. Wei, M., Lin, K., Sun, L.: Shear thickening fluids and their applications. Mater. Des. **216**, 110570 (2022)
10. Ge, J.-q., et al.: Numerical and experimental study on the ultrasonic-assisted soft abrasive flow polishing characteristics. Int. J. Adv. Manuf. Technol. **112**, 3215–3233 (2021)
11. Guo, L., et al.: Shear-thickening polishing of inner raceway surface of bearing and suppression of edge effect. Int. J. Adv. Manuf. Technol. **121**(5–6), 4055–4068 (2022)
12. Li, M., et al.: Origin of material removal mechanism in shear thickening-chemical polishing. Int. J. Mach. Tools Manuf. **170**, 103800 (2021)
13. Li, M., et al.: Evolution and equivalent control law of surface roughness in shear-thickening polishing. Int. J. Mach. Tools Manuf. **108**, 113–126 (2016)
14. Nguyen, D.-N., et al.: Machining parameter optimization in shear thickening polishing of gear surfaces. J. Mater. Res. Technol. **9**(3), 5112–5126 (2020)
15. Shao, Q., et al.: Shear thickening polishing of the concave surface of high-temperature nickel-based alloy turbine blade. J. Mater. Res. Technol. **11**, 72–84 (2021)
16. Fu, K., et al.: Confined compression behaviour of a shear thickening fluid with concentrated submicron particles. Compos. Commun. **10**, 186–189 (2018)
17. Gürgen, S., Li, W., Kuşhan, M.C.: The rheology of shear thickening fluids with various ceramic particle additives. Mater. Des. **104**, 312–319 (2016)
18. Gürgen, S., Sofuoğlu, M.A.: Integration of shear thickening fluid into cutting tools for improved turning operations. J. Manuf. Process. **56**, 1146–1154 (2020)
19. Gürgen, S., Sofuoğlu, M.A., Kuşhan, M.C.: Rheological modeling of multi-phase shear thickening fluid using an intelligent methodology. J. Braz. Soc. Mech. Sci. Eng. **42**, 1–7 (2020)
20. Sun, Z., et al.: Experimental investigations on the magnetorheological shear thickening finishing for sine surface. In: International Conference on Nanomanufacturing. Springer (2021)
21. Wang, S., et al.: Vibration-free surface finish in the milling of a thin-walled cavity part using a corn starch suspension. J. Mater. Process. Technol. **290**, 116980 (2021)
22. Dong, X., et al.: Fine finishing of internal surfaces using cassava starch medium. J. Mater. Process. Technol. **315**, 117918 (2023)
23. Lyu, B., et al.: Experimental study on shear thickening polishing of cemented carbide insert with complex shape. Int. J. Adv. Manuf. Technol. **103**, 585–595 (2019)
24. Nguyen, D.-n.: Simulation and experimental study on polishing of spherical steel by non-Newtonian fluids. Int. J. Adv. Manuf. Technol. **107**(1–2), 763–773 (2020)
25. Kumar, M., Alok, A., Das, M.: Experimental and simulation study of magnetorheological miniature gear-profile polishing (MRMGPP) method using flow restrictor. J. Mech. Sci. Technol. **35**, 5151–5159 (2021)

26. Jiang, W., et al.: Shear-thickening behavior of polymethylmethacrylate particles suspensions in glycerine–water mixtures. Rheol. Acta. **49**, 1157–1163 (2010)
27. James, N.M., et al.: Interparticle hydrogen bonding can elicit shear jamming in dense suspensions. Nat. Mater. **17**(11), 965–970 (2018)
28. Hermes, M., et al.: Unsteady flow and particle migration in dense, non-Brownian suspensions. J. Rheol. **60**(5), 905–916 (2016)
29. Sun, L., et al.: Shear thickening fluid based on silica with neodymium oxide nanoparticles. Bull. Mater. Sci. **43**, 1–6 (2020)
30. Yuan, X., et al.: Forced vibration mechanism and suppression method for thin-walled workpiece milling. Int. J. Mech. Sci. **230**, 107553 (2022)
31. Zhang, Q., et al.: Processing technology and ballistic-resistant mechanism of shear thickening fluid/high-performance fiber-reinforced composites: a review. Compos. Struct. **266**, 113806 (2021)
32. Kim, Y., et al.: High-velocity impact onto a high-frictional fabric treated with adhesive spray coating and shear thickening fluid impregnation. Compos. Part B. **185**, 107742 (2020)
33. Liu, Z., et al.: Novel shear thickening gel/high-performance fiber reinforced flexible composites: multi-velocity impact and functional properties. Polym. Compos. **44**, 3036 (2023)
34. Nagy-György, P., Bene, J., Hős, C.: An optimization-based design approach for a novel self-adjuster using shear thickening fluid. Struct. Multidiscip. Optim. **64**, 4161–4179 (2021)
35. Ding, J., et al., Review on Shear Thickening Fluids and Applications. 2013
36. Brown, E., et al.: Shear thickening and jamming in densely packed suspensions of different particle shapes. Phys. Rev. E. **84**(3), 031408 (2011)
37. Chang, L., Man, Z., Ye, L.: A study on the mechanical polishing technique by using shear thickening fluids. J. Micromech. Mol. Phys. **6**(03), 25–29 (2021)
38. Man, Z., Chang, L.: Shear thickening fluid in surface finishing operations. In: Shear Thickening Fluid: Theory and Applications, pp. 99–114 (2023)
39. Nguyen, D.-N., et al.: Simulation and optimization study on shear thickening polishing of complex surfaces. In: Proceedings of the 2nd Annual International Conference on Material, Machines and Methods for Sustainable Development (MMMS2020). Springer (2021)
40. Zhu, Z.-H., et al.: Fast-tool-servo-controlled shear-thickening micropolishing. Int. J. Mach. Tools Manuf. **184**, 103968 (2023)
41. Sun, P., et al.: Investigation on the performance of fluid jet polishing using shear thickening slurry. In: Optical Manufacturing and Testing XII. SPIE (2018)
42. Preston, F.: The theory and design of plate glass polishing machines. J. Glass Technol. **11**(44), 214–256 (1927)
43. Marinescu, I.D., Uhlmann, E., Doi, T.: Handbook of Lapping and Polishing. CrC Press (2006)
44. Cui, X., et al.: Numerical simulation of shear jamming in a shear thickening fluid under impact. Rheol. Acta. **62**(4), 225–238 (2023)
45. Peters, I.R., Majumdar, S., Jaeger, H.M.: Direct observation of dynamic shear jamming in dense suspensions. Nature. **532**(7598), 214–217 (2016)
46. Ji, S., Xiao, F., Tan, D.: Analytical method for softness abrasive flow field based on discrete phase model. SCIENCE CHINA Technol. Sci. **53**, 2867–2877 (2010)
47. Brown, E., Jaeger, H.M.: Shear thickening in concentrated suspensions: phenomenology, mechanisms and relations to jamming. Rep. Prog. Phys. **77**(4), 046602 (2014)
48. Mai-Duy, N., Phan-Thien, N., Khoo, B.C.: Investigation of particles size effects in dissipative particle dynamics (DPD) modelling of colloidal suspensions. Comput. Phys. Commun. **189**, 37–46 (2015)
49. Kroupa, M., et al.: Utilizing the discrete element method for the modeling of viscosity in concentrated suspensions. Langmuir. **32**(33), 8451–8460 (2016)
50. Chen, K., et al.: A hybrid molecular dynamics study on the non-Newtonian rheological behaviors of shear thickening fluid. J. Colloid Interface Sci. **497**, 378–384 (2017)
51. Wonisch, A., et al.: A comprehensive simulation scheme for tape casting: from flow behavior to anisotropy development. J. Am. Ceram. Soc. **94**(7), 2053–2060 (2011)
52. Kang, M., et al.: Material removal mechanism of non-resonant vibration-assisted magneto-rheological finishing of silicon carbide ceramics. Int. J. Mech. Sci. **242**, 107986 (2023)

Chapter 7
Shear Thickening Materials for High Performance Smart Protection

Ehteshamul Islam, Priyanshi Goel, and Leena Nebhani

7.1 Introduction

The development of smart shear thickening material (STM) has been discussed immensely on various fronts due to their unique time-dependent behavior. Novel STM ensures the safety of personnel or any device against low- or high-velocity threats. STMs have drawn tremendous attention from researchers for understanding methods to enhance the impact resistance of existing materials. STMs can be divided into two categories as shear thickening fluids (STFs) and shear thickening gels (STGs). STFs are concentrated particle suspensions which undergo microstructural rearrangement and show transition from a state of flowable liquid to solid in the event of a sudden impact [1–4]. The non-linear rheological behavior of STF leads to a drastic increase in its viscosity as the applied shear rate increases. In the past, the shear thickening behavior of concentrated suspensions was considered as a drawback for many applications such as paper coating and pumping slurries. However, later its potential as a shock absorber has been realized and used for the development of smart protection systems [5–9]. The nature of STFs depends on the physicochemical nature of the particles, dispersing medium, and volume fraction of particles. STF is usually fabricated by dispersing sub-micron particles into a low molecular weight carrier liquid. In many studies, different types of techniques have been used to disperse the particles in the carrier liquid to get a uniform dispersion. Ultrasonic vibration, mechanical mixing, solvent exchange, homogenization, centrifugal mixing, ball milling, etc. are commonly used methods for STF preparation. Ultrasonic vibration technique is the most reported method after mechanical mixing for getting a uniform dispersion [10–12]. In this method, ultrasonic cavitation is

E. Islam · P. Goel · L. Nebhani (✉)
Department of Materials Science and Engineering, Indian Institute of Technology Delhi, New Delhi, India
e-mail: Leena.Nebhani@mse.iitd.ac.in

S. Gürgen (ed.), *Smart Systems with Shear Thickening Fluid*,
https://doi.org/10.1007/978-3-031-53570-3_7

generated by the sound waves in a low viscosity carrier liquid which generates high pressure jet between the particles. High-pressure jet helps to distribute the particles to obtain a uniform dispersion.

Various rheological mechanisms such as order-disorder and hydrocluster formation have been proposed to understand the shear thickening behavior in the suspensions. Among these mechanisms, formation of hydroclusters is the most accepted approach for explaining the onset of shear thickening in concentrated suspensions. Recently, it has been reported that particle surface interactions with other particles or carrier liquid can affect the state of the particle. According to the recent studies on STFs, the hydrodynamic lubrication forces and frictional forces dictate the shear thickening behavior in a dense suspension of particles [13–18]. The frictional contact forces subside the effect of lubrication forces in differentiating the discontinuous shear thickening from the continuous shear thickening. Contact forces grow in a dense suspension as shear rate increases and ultimately the suspension reaches a complete jammed state [19–21]. STF has been used to impregnate high-performance fabrics for developing an efficient anti-impact system [12, 22–36]. The impact properties of the STF/fabric structures can be controlled by several factors such as particle concentration, particle size, molecular weight of carrier liquid, and additives. However, due to the fluidic nature and storage instability, STF is difficult to be incorporated into wearable fabrics.

As an alternative to STF, recently supramolecular polymer networks with dynamic linkages, usually termed as STGs, have been extensively explored to develop smart protection systems. In STGs, an increase in its storage modulus occurs, leading to phase transition from viscous to elastic state as the angular frequency increases [37–41]. The main reason of this change is originated from the simultaneous dissociation-association of dynamic cross-links between boron and oxygen bonds in STGs [42, 43]. At lower strain rates, these transient cross-links have enough time to break and reform and thereby showing pure viscous behavior. However, at higher strain rates, these cross-links are lack of sufficient time for breaking. As a result, these dynamic bonds do not show a failure and thereby withstanding the applied loadings. This behavior is observed as an instantaneous increase in the storage modulus at high strain rates [44, 45]. During the transition period, STG absorbs the impact energy, which provides excellent anti-impact properties for protective structures. In recent years, due to its unique rate-dependent properties, STG has become a promising material to increase impact energy absorption and dissipate a greater amount of stress in the protective systems [38, 46–49]. As a known fact, STGs behave like a flowing liquid at larger time scales; therefore, integrating STGs into a dimensionally stable structure is required to introduce new smart protecting gears for diverse engineering applications. Unlike STF, STG is a viscoelastic material with higher initial viscosity and shows cold flow at rest. Therefore, it is easier to store STG without particle sedimentation experienced with STFs. The viscoelastic properties of polyborosiloxane (PBS), which is the main component in STGs, is dependent on the molecular weight. Seetapan et al. [50]

prepared the physically cross-linked PBS by mixing hydroxy terminated PDMS and boric acid at 120 °C for 48 h. When hydroxyl group reacts with boric acid, a reversible cross-linked complex in the form of Si-O-B bonds is formed. Various boron sources such as boric acid, trialkyl borate, and borohydride have been used for the synthesis of PBS. Since STG shows strain-induced phase change from viscous to elastic state, its rheological behavior can be correlated with the change in its moduli. In previous studies, oscillatory shear rheology is a choice to characterize the STG moduli over a broad range of applied shear strains.

The potential of STG as an anti-impact material has been extensively explored by researchers. Owing to STG's strain-dependent behavior, various anti-impact products have been prepared for ballistics, wearable electronics, and sports applications [51–53]. Unlike other composite structures, STG-incorporated sandwich panels show uniform distribution of stress generated during high velocity impact [54]. Tang et al. [55] prepared an STG by heating the mixture of boric acid and hydroxy-terminated PDMS (1:5 weight ratio) at a temperature of 160–180 °C (Fig. 7.1a). Due to the electron deficient p-orbital of boron, it forms cross-link with the electron-rich oxygen atoms. Various possible structures of these cross-links are reported as shown in Fig. 7.1b. Liu et al. [57] incorporated the STG into polyurethane foam with a one-step foaming method. They developed a single-pot method to prepare STG/polyurethane foam, which shows improved energy absorption. Researchers found that STGs can greatly enhance the ballistic performance of bulletproof jackets without compromising the comfort to the wearers [58–60]. Recently, the

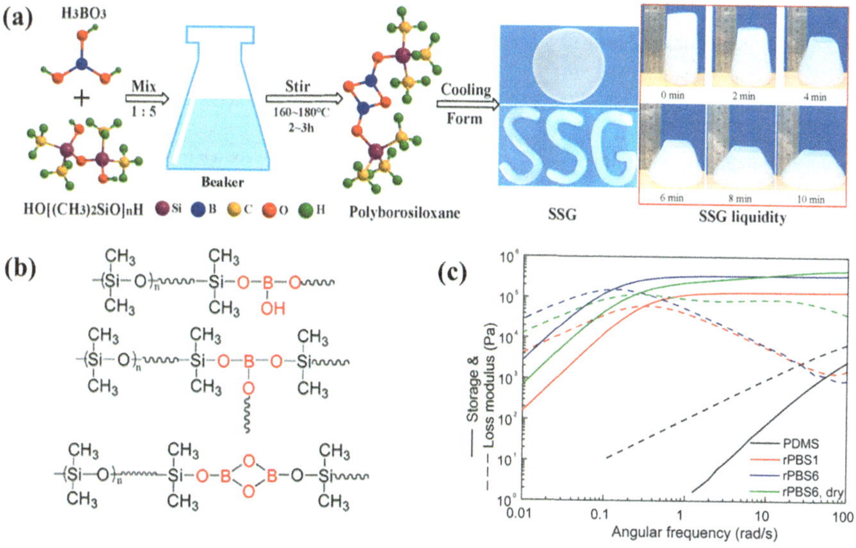

Fig. 7.1 (a) STG synthesis [55], (b) possible structure of STG [48], (c) storage and loss moduli of PBSs for different reaction times [56]

commercial grade, D3O developed STG-impregnated polyurethane products for multiple applications ranging from bulletproof vests to sports gears. Interestingly, due to its rate-dependent behavior and self-healing properties, STGs have been used to develop stretchable devices. Due to excellent shear stiffening capabilities, STG can be seen as a magneto-rheological material. For example, Wang et al. [61] prepared a conductive STG/Kevlar fabric which exhibits excellent shear stiffening effect under applied magnetic field.

7.2 Shear Thickening Materials and Their Composites

STMs generally consist of two kinds of materials which are STF and STG. They differ in their composition, state, and thickening mechanism. These smart materials have been widely investigated for their fast responsiveness toward impact absorption against various degrees of loadings. Table 7.1 shows the chemical compositions for different STMs.

Table 7.1 Chemical compositions for different STMs

Type of STM	Main phase	Dispersing medium	Particle concentration	References
STF	Silica (1.92 μm)	Glycerol/water	86 wt%	[62]
	Fumed silica (12 nm)	PPG (725 g/Mol)	3–10 wt%	[63]
	Colloidal silica (450 nm)	PEG (200 g/Mol)	52 vol%	[64]
	Modified fumed silica (12 nm)	PPG (400 and 2000 g/Mol)	10.3 wt%	[65]
	Fumed silica (12 nm)	PPG (1000 g/mol)	67–68 wt%	[66]
	Precipitated calcium carbonate (30–200 nm)	Glycerin	37–54 vol%	[67]
	Colloidal silica (100 nm)	PEG (200 g/Mol)	60 wt%	[68]
	Colloidal silica (220 nm)	PEG (400 g/Mol)	35–58 vol%	[69]
	Colloidal silica/cellulose bead	PEG (200 g/Mol)	62 wt%	[70]
	PSt-PHEMA (500 nm)	Ethylene glycol	50 wt%	[71]
	Colloidal silica (500 nm)	PEG	15, 25, 35, and 45 wt%	[72]
	Fumed silica (20 nm)	PEG (400 g/Mol)	20 wt%	[73]
	Modified silica (675–710 nm)	PEG (200 g/Mol)	40–50 wt%	[74]
	Colloidal silica (500 nm)	Ionic liquids	50–64 wt%	[75]
	Colloidal silica (300 nm)	PEG (400 g/Mol)	70 wt%	[76]
	Fumed silica (200–300 nm)	PPG	15, 18, and 20 wt%	[77]
	Silica (370 μm)	PEG (400 g/Mol)	57 vol%	[78]

(continued)

Table 7.1 (continued)

Type of STM	Main phase	Dispersing medium	Particle concentration	References
STG	Hydroxy terminated PDMS, boric acid	–	–	[50]
	Trimethylsiloxy terminated PDMS, boric acid, hexane	–	–	[79]
	Dichlorfenylmethylsilane, hydroxy silicone oil, sodium tetraborate, water	–	–	[80]
	Dimethyl siloxane, boric acid, acetone, and benzoyl peroxide	–	–	[81]
	Hydroxy terminated PDMS, boric acid, methanol	–	–	[82]
	Dimethyl siloxane, boric acid, ethanol, and benzoyl peroxide	–	–	[83]
	Dimethylsiloxane, boric acid	–	–	[84]
	PDMS, boric acid, ethanol, benzoyl peroxide, carbonyl iron powder	–	–	[85]
	Hydroxy terminated PDMS, boric acid	–	–	[55]
	PDMS, boric acid	–	–	[58]

7.2.1 *Shear Thickening Fluid*

STF is a colloidal suspension comprising dispersed micro and nanoparticles in the carrier liquid. The particles such as silica, polymethylmethacrylate, cornstarch, calcium carbonate, or titanium oxide are suspended in inert liquids such as water, ethylene glycol (EG), and polyethylene glycol (PEG) to obtain non-Newtonian fluids, which exhibit exponential increase in viscosity by a few orders of magnitude as increasing shear rate or shear stress. The deformation is reversible and thus, it reverts to the initial fluidic state once the external force is removed. STF performance is closely related to the dispersed particles and carrier liquids. It is well known that physicochemical parameters related to dispersing particles such as hardness, roughness, aspect ratio, size distribution, solid loading, surface chemistry, etc. significantly affect the rheological behavior of STFs [86].

On account of the impressive properties of these fluids, STF-impregnated fabrics and composites have been widely used for impact-absorbing applications. However, STF applications have some shortcomings. The sedimentation of suspended particles and water absorption after prolonged exposure to air due to hygroscopic nature of the dispersed medium results in time-dependent deterioration in the fluid's properties. In addition, the low viscosity of STF at rest state makes this fluid difficult to encapsulate in the composites and inevitably losing its efficiency in protection. To overcome these kinds of drawbacks and maintain the stability and shear thickening effect in STF, various modifications have been made for promoting it to play a greater role in practical applications [87].

7.2.2 Shear Thickening Gel

STG is another kind of STMs attracting the attention of researchers around the globe. Also known as STG, it is a viscoelastic polymer generally cross-linked lightly with dynamic boron-oxygen bonds, which allow the material to absorb impact energy by showing a phase transition from viscous liquid to rubbery state followed by a glassy state under shear [88]. STG exhibits low modulus at unloaded conditions. However, external loading initiates a jump in storage modulus, elastic modulus, and yield stress, leading to a sudden stiffening effect. STGs are often synthesized by treating polydimethylsiloxane or hydroxy terminated silicone oil with boric acid and benzoyl peroxide (BPO) at high temperatures, forming B-O bonds mainly responsible for the shear thickening effect [89, 90]. At low strain rates, the entanglements in the molecular network can be detangled through internal mobility in the molecular chains. The chains undergo faster relaxation due to exchange of dynamic B-O bonds, leading to rubbery state. However, at high strain rates, the loading rate is higher than the rate of association and dissociation of dynamic bonds, leading to severe molecular chain entanglement. Consequently, large numbers of B-O cross bonds are broken to endure the external stimuli and to dissipate impact energy, which results in the significantly increased storage modulus [91].

In contrast to STF, STG possesses lesser cross-linking density providing higher initial viscosity and mechanical properties. Absence of any suspended particles in the liquid phase makes STG more stable against issues like particle sedimentation and solvent volatilization [87]. STG reinforcement in STF/fabric composites not only increases the stability of the material against moisture absorption, but also increases the ballistic properties by limiting the sliding of the yarns and even distribution of impact forces around the composite [88]. Despite the better mechanical properties than STF, STG exhibits cold flow behavior at rest state, which hinders its practical utilization. This is a time-dependent irreversible deformation, leading to the collapse of polymer matrix with increasing time frame. However, encapsulation of STG in porous foams can be a potential solution [89]. The advanced analysis for the energy absorption and rheological behavior of STG has enabled the development of an array of STG-reinforced composites regarding anti-impact devices for military and sports applications during impulse loading.

7.2.3 Composites Based on Shear Thickening Materials

Early 2000s has been marked for the development of technology based on STF across the world. British engineer Richard Palmer commercialized the energy-absorbing material based on STM under the name of D3O® [92]. Simultaneously in the other part of the world, the U.S. Army Research Laboratory and the University of Delaware developed an STF, to create a new generation of military battle suit [93]. A series of high-performance composites integrating the fast responsiveness of STM toward the strain impacts and exceptional mechanical properties of advanced 2D [94, 95] and 3D warp-knitted spacer fabrics [96] have been developed for the anti-impact application in military and sports equipment.

Apart from these fabrics, polymers like ethylene vinyl acetate foam have also been reinforced with STG. These materials have enhanced elastic modulus and compressive strength at high strain rates and improved toughness attributed to the reinforcing effect of STG. Improved storage and loss moduli of the composites enable effective dissipation of energy through viscous behavior. Hence, these factors make the foam resilient and stiffer, demonstrating the use of composite in human protective equipment such as buffer layer which is placed behind the ballistic layers [97, 98].

7.2.4 Fabrication Techniques of Shear Thickening Material-Based Composites

Polymer composites based on STM have revolutionized the world of armor and other protection gear. Substituting heavy materials like metallic alloys and ceramic composites with STM-impregnated multilayer fabrics brings lightness and relief for the body movements. Along with compositions in the protective structures, fabrication techniques also play a vital role in optimizing the balance between impact properties and flexibility of the materials. Various techniques have been explored by the researchers to incorporate STM into the protective fabrics. Broadly dip and dry, hand lay-up, vacuum infusion, and padding methods are used for fabric impregnation as depicted in Fig. 7.2.

High viscosity of STM impedes the complete impregnation of fabrics. For the efficient immersion of fabric layers into the viscous material, STF is diluted using solvents such as ethanol or acetone, converting it into an easily dispersive liquid for uniform impregnation. After that, fabrics are soaked into the solution for a certain period of time, and dried in an oven until the solvent is removed. The dip-and-dry method provides the uniform distribution of fluid or gel into the gaps between the surface and the layers of fabrics along with controlling over the resin mass. On the other hand, taking care of minute difference of impregnation, which includes direct placement of STM over the fabric instead of dipping into the solution, the method is known as hand lay-up method [88]. The fluid infusion with extremely low viscosity or no dilution using vacuum infusion process leads to bubble-free and lighter composites along with enhanced anti-impact properties [101]. Due to the vacuum environment, the fluid infuses into the fabrics and the process is conducted until a complete impregnation is achieved. STMs have been widely incorporated into different fabrics using these techniques to develop protection gears for low and high velocity impact applications [88, 102, 103].

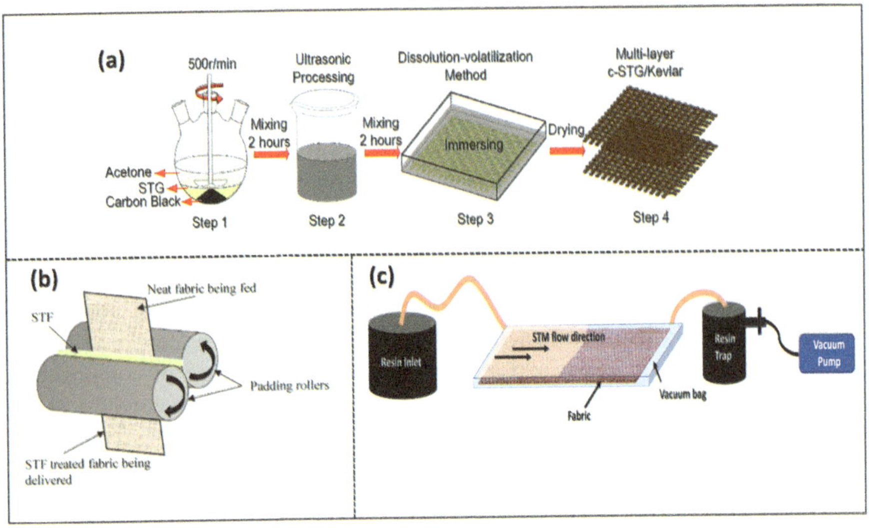

Fig. 7.2 Fabric impregnation by (**a**) dip-and-dry method [99], (**b**) padding rollers [100], and (**c**) vacuum infusion process

All these methods are restricted to a small-batch size production and consume comparatively more volume of STM than another technique, called padding. Impregnation in padding method includes dipping of fabric layers into the diluted solution followed by compressing it through two rollers under a certain pressure. Some studies were conducted to investigate the effect of padding pressure and consequent STF add-on, and found that higher padding pressure squeezes out the fluid from the fabrics, leading to less STF add-on. A strong conclusion on effect of increasing padding pressure relies on the type of fabrics used. In general, higher inter-yarn friction provides better impact properties for loose weaved fabrics while the increase in friction generates localized stress concentration, leading to deterioration in impact resistance performance for tightly woven fabrics [103, 104]. Padding not only enhances the performance of the composites, but also opens a new technique for continuous manufacturing [103–106].

7.3 Impact Properties of Shear Thickening Material-Impregnated Fabrics

Owing to the thickening characteristics in STMs, these intelligent materials have been widely used in anti-impact applications, such as protective gears and energy-absorbing applications. Impact behavior of STM-based composites has been investigated under quasi-static, low velocity, medium velocity, and high velocity conditions. Figure 7.3 shows various testing methods for different impact conditions.

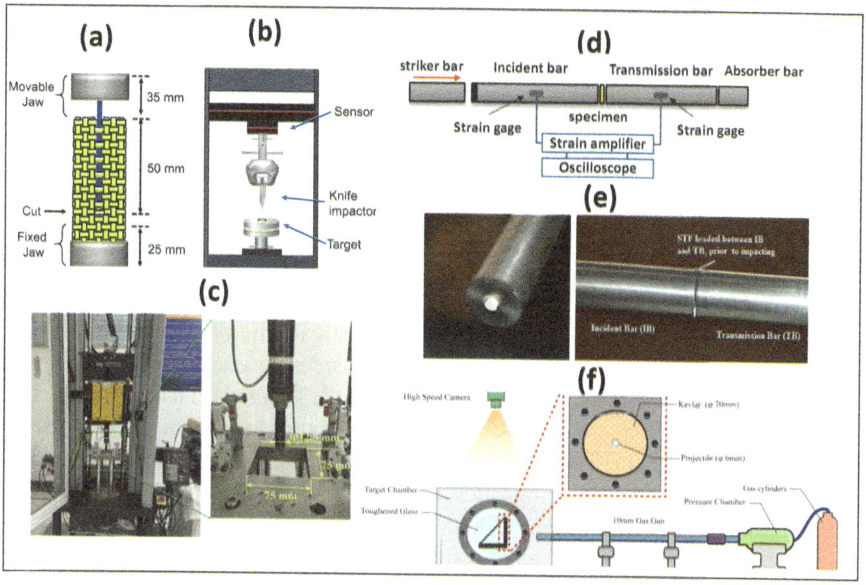

Fig. 7.3 (**a**) Yarn pull-out testing, (**b**) quasi-static stab testing [107], (**c**) drop weight testing [108], (**d**) split Hopkinson pressure bar testing [88], (**e**) specimen loaded between the incident and transmission bars [109] and (**f**) ballistic impact testing [110]

7.3.1 *Quasi-Static Impact*

Impact damages like puncture, spike, or knife stab can be successfully investigated through the quasi-static testing. A variety of high-performance fabrics are available for developing puncture or stab-resistant structures [111]. However, multiple layering of fabrics leads to bulky and rigid composites, bothering users with uncomfortable texture. Resin infusion in fabrics leads to enhanced adhesion between yarns, thus improving the puncture and stab resistance to certain values but at the cost of flexibility. On the other hand, fabric impregnation with STM leads to a significant decrease in penetration depth as well as increasing inter-yarn adhesion. It is obvious that STM application to protective fabrics yields enhanced impact resistance without compromising the flexibility in structures [112]. Despite very small shear rates in quasi-static testing, thickening mechanism takes place in STMs. It is proven that thickening onset corresponds to low shear rates, where small shear rates are sufficient for viscosity increase [94]. The penetration resistance performance of STF-treated fabrics revealed that impregnated fabrics not only bear higher loading than neat fabrics, but also show different modes of failure. STF-impregnated fabrics fail due to yarn and fiber breakage while neat fabrics show yarn pull-out during the deformation. The difference in failure modes is attributed to the increased adhesion between the fibers caused by the STF. Supportively, higher forces are required to pull-out a yarn from STF-impregnated fabrics, which proves that impregnated fabrics have a good potential for developing protective systems [107, 113, 114].

Hardness of dispersed phase plays an important role in stab resistance. Adhesion within fabrics, frictional interactions between yarns, and hard particle depositions in yarn gaps get more importance during stab conditions regarding the yarn breakage [115].

7.3.2 Low Velocity Impact

Impact resistance at low velocity conditions can be investigated by using drop weight testing [116]. Researchers generally investigate the protective behavior of composites against low velocity impacts by considering energy absorptions and maximum resistance forces following the ASTM D7136 standard using drop weight testing. Impact velocity greatly affects the mechanical properties of the composites [117]. Zhang et al. [118] studied STF/impregnated ultra high molecular weight polyethylene (UHMWPE) composites under low velocity impacts by using the velocities at 2.5, 3.0, and 3.5 m/s. They concluded that shear thickening effect is pronounced at higher impact velocities, thereby enhancing protective performance of the composites. Tan et al. [108] investigated an STG-based structure by varying the impact velocities from 5 to 7 m/s. According to their study, energy-absorbing capabilities are greatly increased by the contribution of thickening properties in STG. On the other hand, impactor geometries have also a significant effect on the impact response of STG-based composites. Striking with a flat-head geometry leads to a larger thickening on the composites while point-contact during impacts as such in cone-shaped impactors restrict the stiffening area on the structures, consequently limiting the STM contribution to the energy-absorbing mechanism [119]. The effect of STM has been well appreciated for enhancing the protective properties of different fiber-reinforced composites; however, the properties are greatly affected by the type of fabrics. Based on a comparison between the STF-treated basalt, carbon, and glass fabrics, glass fabrics show higher jump in energy absorbing due to STF impregnation. From this study, it can be stated that the coupling between STF and glass fibers results in a high-performance composite against low-velocity impact conditions.

7.3.3 Intermediate and High Velocity Impact

Composites based on STMs serve to stop intermediate and high velocity projectiles by absorbing their kinetic energies. Advanced fibers like aramid, polyethylene, and carbon fibers impregnated with STM have been widely used to develop bulletproof armors and other protective equipment. STF impregnation enhances the ballistic limit and energy-absorption capabilities of protective fabrics as well. STF-treated fabrics stop the projectiles even at low mass fraction of STF contents [110]. The impregnation leads to an increase in inter-yarn frictions within fabrics, thereby providing higher energy-absorption performance. This turns out to be one of the critical factors along with shear thickening rheology to enhance the impact resistance of composites [110, 120].

Conventional rheology tests cannot provide sufficient data about the high-velocity impact response of STM since the deformation at high-velocity impacts takes place in a very small time period. The Split Hopkinson Pressure Bar (SHPB) testing can effectively be used to study the dynamic properties of the polymers under high strain rate varying from 10^2 to $10^4\,s^{-1}$ [95, 109, 121]. The system includes three bars: Incident Bar (IB), Transmission Bar (TB), and Striker Bar (SB). The sample is loaded at the interface of IB and TB. During the testing, the SB hits the IB, and a fraction of the loading pulse reflects through the IB while the remaining is transmitted through the TB. The sample has shear thickening transition, showing peak stresses depending on the strain rates [109]. Moreover, SHPB testing provides the evaluation of characteristic transition time, which is another critical parameter for ballistic applications.

Furthermore, another crucial analytical technique for the evaluation of these composites under high-velocity impact involves ballistic testing. This test includes a gas gun for sending projectiles to the targets with different velocities. Initial and residual velocities of projectiles are measured to find the energy absorbed by the targets [105]. High velocity impact response is governed by the combined behavior of fabric layers and incorporated STM into the structure. Energy-absorption rate in yarns and STM behaves differently by increasing impact velocity. Loading with high strain rates does not provide a sufficient time for the impacted yarns to be pulled out in the woven structure and thereby straining the yarns immediately. On the other hand, high strain rates create a sudden thickening in STM texture. For this reason, energy-absorbing capabilities in the yarns and STM are significantly changed depending on the shear rates.

7.4 Shear Thickening and Impact Mitigation Mechanism

STGs are fundamentally siloxane-boron type of polymers, and the thickening effect in STGs originates from the supramolecular interactions such as hydrogen bonds, and dative bonds between boron and oxygen in the Si-O-B bonds. This interaction occurs due to the transfer of electrons from the electron-rich oxygen to the empty p-orbital of the boron atom. This mechanism leads to the formation of weak dative bonds between them. Due to the formation of these cross-links through dative bonds, shear-induced phase transformation in STGs occurs, which results in a significant increase in storage modulus. It has been reported that, at low strain rates, the molecular chains in STGs have a sufficient time to relax and get disentangled. The mechanical properties of STGs are governed by simultaneous breakage and reformation of transient B-O bonds, which allow the molecular chains to have internal mobility. However, at higher strain rates, the timescale for molecular motions becomes lesser than the characteristic relaxation time of B-O bonds. As a result,

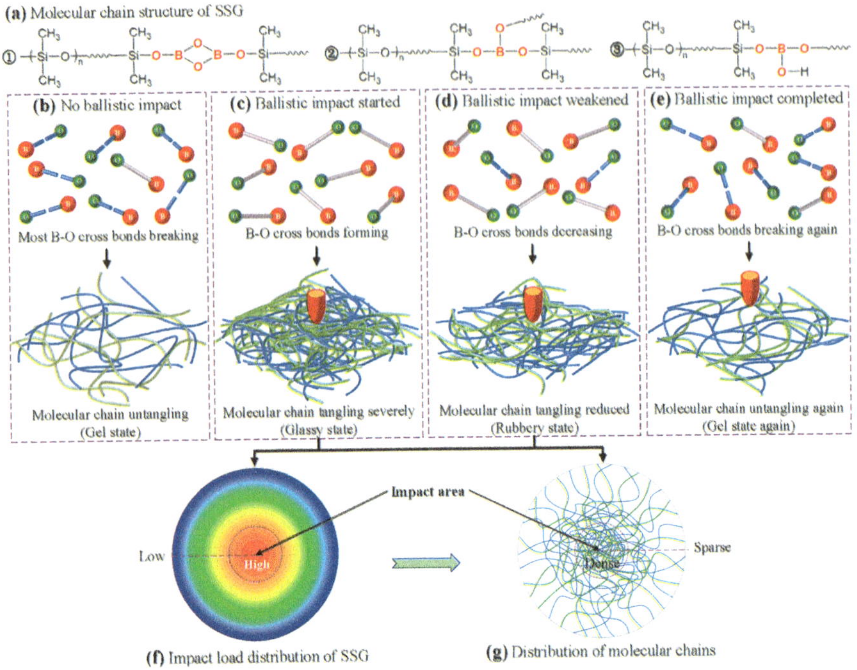

Fig. 7.4 Thickening mechanism in STG under ballistic conditions [98]

large number of B-O bonds is broken, thereby generating more resistance against the movement of molecular chains, which leads to a high stiffness in the polymer. Subsequently, B-O bonds are broken, thereby dissipating impact energy with an increasing storage modulus.

According to Tang et al. [98], impact energy dissipation occurs in STG through four distinct phases (Fig. 7.4). At a low shear rate, STG is in the gel state, where the breaking velocity of B-O bonds is greater than the deformation. As a result, most of the B-O bonds are broken, and the molecular chains get disentangled. It is known as a weak jamming effect because of lower storage modulus. STG transforms into a glassy state during an impact. At this stage, breaking velocity of B-O bonds is lower than the deformation rate of STG, and thus B-O bonds do not have sufficient time to break, and the molecular chains get entangled, enhancing the jamming effect. Therefore, the motion of molecular chains gets restricted, causing an increase in storage modulus. As the stress waves diminish in the material, STG reaches a rubbery state. In this state, B-O bonds show breaking, and the entanglement density decreases with weakening jamming. Upon completely dissipating the stress waves, molecular chains reach the most stable state, where the molecular chains are entangled, causing the material to get back to the initial gel texture (Fig. 7.4b–e).

Tang et al. [98] correlated the impact load distribution with the molecular entanglement distribution at ballistic conditions. It is shown that ballistic impact attenuates from the point of contact region toward the surrounding area on the target

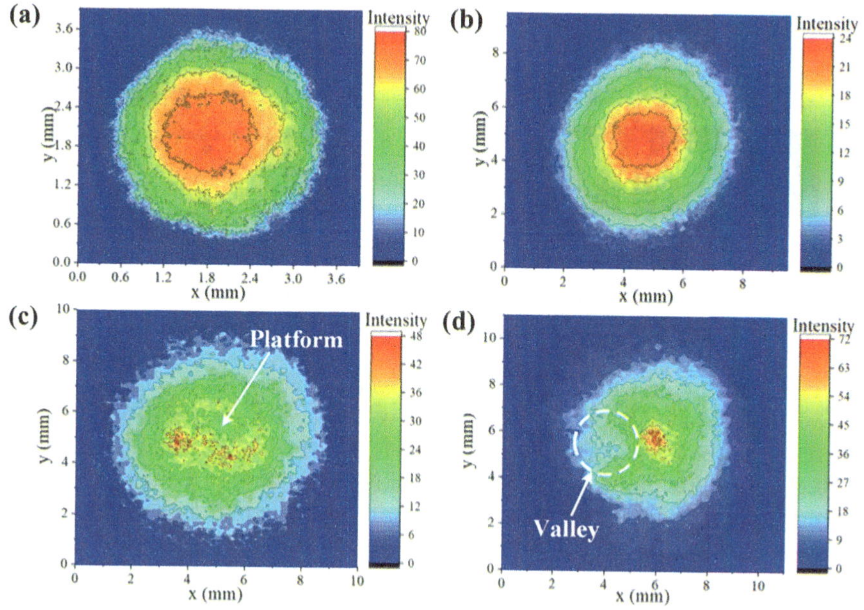

Fig. 7.5 Illumination-intensity distribution graph generated during hammer impact testing of (**a**) stress-activated mechanoluminescence device from a height of 0.1 m (**b**) stress-activated mechanoluminescence/STG from a height of 0.1 m, (**c**) 0.2 m, (**d**) 0.3 m [53]

surface. According to the distribution of loading in STG, the molecular chains show a denser entanglement at the center (Fig. 7.4f and g). Therefore, the impact-energy dissipation in STG occurs through stress distribution from the higher entanglement region to the lower entanglement one.

Another method of assessing the impact-energy dissipation in STG was recently reported and characterized by mechanoluminescence method. Due to the mechanoluminescence property of manganese-doped zinc sulphide (ZnS:Mn) particles, it is utilized for the measurement of energy absorption behavior. In this method, a stress-activated machanoluminescence device is placed below the STG and measures the instantaneous intensity at different drop heights, simultaneously the camera, which is placed below the setup, records the images of illumination. The illumination-intensity distribution graph generated in hammer impact testing is shown in Fig. 7.5. Zang et al. [53] found that the drop height of hammer shows an additional effect on force and instantaneous intensity signals during impact. It was observed that STG introduces low strain at a lower drop height (0.1 m), and the force initially increases, then gradually decreases. As the drop height increases to 0.2 m and 0.3 m, STG shows a transition from viscous liquid state to rubbery state due to higher strain. Shear thickening effect, which is observed at higher drop heights, can be assigned to the generation of second peak followed by the yield point. This shows that the dynamic strength of STG significantly enhances as the strain rate increases. The stress-activated mechanoluminescence device shows a concentric illumination

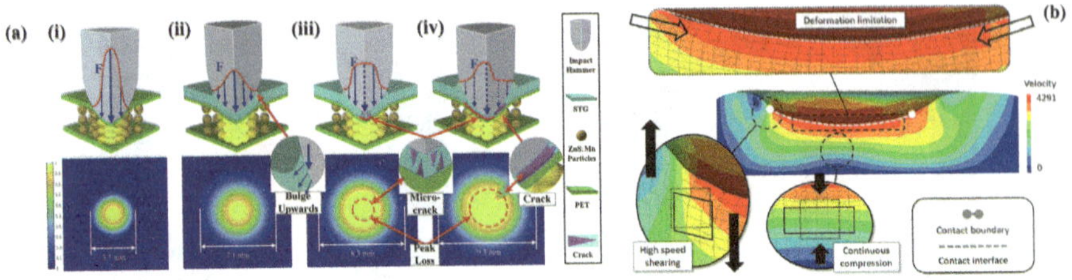

Fig. 7.6 (a) Impact energy dissipation mechanism under impact loading [53], and (b) deformations in STG at impact velocity of 5 m/s [123]

intensity distribution with a peak intensity at the center, which is 35% of the total impact area. Moreover, the impacted area is measured as 3.1 mm in diameter. Interestingly, the impact diameter is extended to 7.1 mm, which is simply associated with the reduction in the energy density and illumination distribution. It is worth noting that the illumination distribution diminishes sharply at the center while remaining unaffected at the edges as drop height increases.

Fan et al. [122] incorporated an STG into sutural structures. Both experimental and numerical simulations were carried out to investigate these composites. They found that structural parameters such as tooth angle at the sutural interface can significantly affect the impact energy dissipation in the composites. Through the mechanoluminescence method, the variations in the light intensity and its distribution were analyzed during the impacts. Internal cracks in STG help dissipating the impact energy, where the illumination intensity gradually fades out (Fig. 7.6a). Ying et al. [123] investigated the ability of STG regarding the stress wave distribution under low velocity impacts. Figure 7.6b shows the deformations in STG based on the numerical simulations at impact velocity of 5 m/s. It is observed that the maximum stress is located at the impact point on the STG. Stress maximization and stress distribution in viscoelastic materials depend on two main factors: (i) velocity gradient and (ii) deformation time.

7.5 Conclusions

In this chapter, shear thickening properties of STMs are discussed. Followed by thorough discussion about the STM-based composites, fabrication techniques are given in detail. The impact response of STM-based structures is discussed for low and high velocity impact conditions. Several studies reveal that STM incorporation into sandwich structures has a great effect on impact energy absorption, hence paving the way for the researchers to design smart protective systems. Despite this fact, in-depth analyses are needed for better understanding the shear thickening mechanism and its correlation with the protective performance of STMs. The thickening characteristics in STMs make these materials an excellent choice for developing smart protective systems.

References

1. Laun, H.M.: Rheological properties of aqueous polymer dispersions. Angew Makromol Chem. **123–124**, 335–359 (1984)
2. Barnes, H.A.: Shear-thickening ("Dilatancy") in suspensions of nonaggregating solid particles dispersed in Newtonian liquids. J. Rheol. (N Y N Y). **33**(2), 329–366 (1989 Feb)
3. Fagan, M.E., Zukoski, C.F.: The rheology of charge stabilized silica suspensions. J. Rheol. (N Y N Y). **41**(2), 373–397 (1997)
4. Lee, J.D., So, J.H., Yang, S.M.: Rheological behavior and stability of concentrated silica suspensions. J. Rheol. (N Y N Y). **43**(5), 1117–1140 (1999)
5. Lee, Y., Wagner, N.: Dynamic properties of shear thickening colloidal suspensions. Rheol. Acta. **42**, 199–208 (2003)
6. Zeng, X.S., Shim, V.P.W., Tan, V.B.C.: Influence of boundary conditions on the ballistic performance of high-strength fabric targets. Int. J. Impact Eng. **32**(1–4), 631–642 (2005)
7. Mahfuz, H., Clements, F., Rangari, V., Dhanak, V., Beamson, G.: Enhanced stab resistance of armor composites with functionalized silica nanoparticles. J. Appl. Phys. **105**(6) (2009)
8. Kalman, D.P., Merrill, R.L., Wagner, N.J., Wetzel, E.D.: Effect of particle hardness on the penetration behavior of fabrics intercalated with dry particles and concentrated particle-fluid suspensions. ACS Appl. Mater. Interfaces. **1**(11), 2602–2612 (2009)
9. Kang, T.J., Hong, K.H., Yoo, M.R.: Preparation and properties of fumed silica/Kevlar composite fabrics for application of stab resistant material. Fibers Polym. **11**(5), 719–724 (2010)
10. Güler, E.S.: Rheological behaviours of silica/water, silica/PEG systems and mechanical properties of shear thickening fluid impregnated Kevlar composites. Bull. Mater. Sci. **41**(4) (2018)
11. Katiyar, A., Nandi, T., Katiyar, P.: Energy absorption of graphene and CNT infused hybrid shear thickening fluid embedded textile fabrics. J. Polym. Res. **28**(9), 1–13 (2021)
12. Huang, C.H., Chien, C.H., Shiu, B.C., Chen, Y.S., Lin, J.H., Lou, C.W.: Reinforcement effects of shear thickening fluid over mechanical properties of nonwoven fabrics. Polymers (Basel). **14**(22), 4816 (2022)
13. Brown, E., Jaeger, H.M.: The role of dilation and confining stresses in shear thickening of dense suspensions. J. Rheol. (N Y N Y). **56**(4), 875 (2012)
14. Heussinger, C.: Shear thickening in granular suspensions: Interparticle friction and dynamically correlated clusters. Phys. Rev. E Stat. Nonlinear Soft Matter Phys. **88**(5), 22–24 (2013)
15. Boyer, F., Guazzelli, É., Pouliquen, O.: Unifying suspension and granular rheology. Phys. Rev. Lett. **107**(18), 1–5 (2011)
16. Wyart, M., Cates, M.E.: Discontinuous shear thickening without inertia in dense non-brownian suspensions. Phys. Rev. Lett. **112**(9), 1–5 (2014)
17. Hermes, M., Guy, B.M., Poon, W.C.K., Poy, G., Cates, M.E., Wyart, M.: Unsteady flow and particle migration in dense, non-Brownian suspensions. J. Rheol. (N Y N Y). **60**(5), 905–916 (2016)
18. Kawasaki, T., Berthier, L.: Discontinuous shear thickening in Brownian suspensions. Phys. Rev. E. **98**(1), 1–15 (2018)
19. Mari, R., Seto, R., Morris, J.F., Denn, M.M.: Shear thickening, frictionless and frictional rheologies in non-Brownian suspensions. J. Rheol. (N Y N Y). **58**(6), 1693–1724 (2014)
20. Mari, R., Seto, R., Morris, J.F., Denn, M.M.: Discontinuous shear thickening in Brownian suspensions by dynamic simulation. Proc. Natl. Acad. Sci. USA. **112**(50), 15326–15330 (2015)
21. Lin, N.Y.C., Guy, B.M., Hermes, M., Ness, C., Sun, J., Poon, W.C.K., et al.: Hydrodynamic and contact contributions to continuous shear thickening in colloidal suspensions. Phys. Rev. Lett. **115**(22), 1–5 (2015)
22. Bai, R., Ma, Y., Lei, Z., Feng, Y., Liu, C.: Energy analysis of fabric impregnated by shear thickening fluid in yarn pullout test. Compos. Part B Eng. **174**, 174(June) (2019)
23. Liu, L., Yang, Z., Zhao, Z., Liu, X., Chen, W.: The influences of rheological property on the impact performance of kevlar fabrics impregnated with SiO2/PEG shear thickening fluid. Thin-Walled Struct. **151**(29), 106717 (2020)

24. Gürgen, S., Sofuoğlu, M.A.: Integration of shear thickening fluid into cutting tools for improved turning operations. J. Manuf. Process. **56**(May), 1146–1154 (2020)
25. Mawkhlieng, U., Majumdar, A., Bhattacharjee, D.: Graphene reinforced multiphase shear thickening fluid for augmenting low velocity ballistic resistance. Fibers Polym. **22**(1), 213–221 (2021)
26. Wetzel, E.D.: The effect of rheological parameters on the ballistic properties of shear thickening fluid (STF)-Kevlar composites. In: AIP Conference Proceedings, pp. 288–293. AIP (2004)
27. Houghton, J.M., Schiffman, B. a., Kalman, D.P., Wetzel, E.D., Wagner, N.J.: Hypodermic needle puncture of shear thickening fluid (STF)-treated fabrics. Proceedings of SAMPE. 2007;(June):3–7
28. Rosen, B.a., Laufer, C.H.N., Kalman, D.P., Wetzel, E.D., Norman, J.: Multi-Threat Performance of Kaolin-Based Shear Thickening Fluid (Stf) -Treated Fabrics. Most. 2007;(June):1–11
29. Zhang, X.Z., Li, W.H., Gong, X.L.: The rheology of shear thickening fluid (STF) and the dynamic performance of an STF-filled damper. Smart Mater. Struct. **17**(3), 035027 (2008)
30. Hassan, T.A., Rangari, V.K., Jeelani, S.: Synthesis, processing and characterization of shear thickening fluid (STF) impregnated fabric composites. Mater. Sci. Eng. A. **527**(12), 2892–2899 (2010)
31. Sun, L.L., Xiong, D.S., Xu, C.Y.: Application of shear thickening fluid in ultra high molecular weight polyethylene fabric. J. Appl. Polym. Sci. **129**(4), 1922–1928 (2013)
32. Kordani, N., Vanini, A.S.: Optimizing the ethanol content of shear thickening fluid/fabric composites under impact loading. J. Mech. Sci. Technol. **28**(2), 663–667 (2014)
33. Cwalina, C.D., Dombrowski, R.D., McCutcheon, C.J., Christiansen, E.L., Wagner, N.J.: MMOD puncture resistance of EVA suits with shear thickening fluid (STF) – Armortm absorber layers. Procedia Eng. **103**, 97–104 (2015)
34. Galindo-Rosales, F.J., Martínez-Aranda, S., Campo-Deaño, L.: CorkSTFμfluidics – a novel concept for the development of eco-friendly light-weight energy absorbing composites. Mater. Des. **82**, 326–334 (2015)
35. Na, W., Ahn, H., Han, S., Harrison, P., Park, J.K., Jeong, E., et al.: Shear behavior of a shear thickening fluid-impregnated aramid fabrics at high shear rate. Compos. Part B Eng. **97**, 162–175 (2016)
36. Majumdar, A., Laha, A., Bhattacharjee, D., Biswas, I.: Tuning the structure of 3D woven aramid fabrics reinforced with shear thickening fluid for developing soft body Armour. Compos. Struct. **178**, 415–425 (2017)
37. Tian, T.F., Li, W.H., Ding, J., Alici, G., Du, H.: Study of shear-stiffened elastomers. Smart Mater. Struct. **21**(12), 125009 (2012)
38. Jiang, W., Gong, X., Wang, S., Chen, Q., Zhou, H., Jiang, W., et al.: Strain rate-induced phase transitions in an impact-hardening polymer composite. Appl. Phys. Lett. **104**(12) (2014)
39. Xu, K., Tan, Y., Xin, J.H., Liu, Y., Lu, C., Deng, Y., et al.: A novel impact hardening polymer with negative Poisson's ratio for impact protection. Mater. Today Commun. **5**, 50–59 (2015)
40. Tang, M., Wang, W., Xu, D., Wang, Z.: Synthesis of structure-controlled polyborosiloxanes and investigation on their viscoelastic response to molecular mass of polydimethylsiloxane triggered by both chemical and physical interactions. Ind. Eng. Chem. Res. **55**(49), 12582–12589 (2016)
41. Tang, M., Zheng, P., Wang, K., Qin, Y., Jiang, Y., Cheng, Y., et al.: Autonomous self-healing, self-adhesive, highly conductive composites based on a silver-filled polyborosiloxane/polydimethylsiloxane double-network elastomer. J. Mater. Chem. A. Mater. **7**(48), 27278–27288 (2019)
42. Chang, Z., He, Y., Deng, H., Li, X., Wu, S., Qiao, Y., et al.: A multifunctional silly-putty nanocomposite spontaneously repairs cathode composite for advanced Li–S batteries. Adv. Funct. Mater. **28**(50), 1–10 (2018)
43. Qi, W., Xiong, H., Peng, Y., Yang, Y., Kang, J., Huang, G., Ren, X., JW.: Highly stretchable and self-healing "solid–liquid" elastomer with strain-rate sensing capability.Pdf. ACS Appl. Mater. Interfaces. **11**(21), 19534–19540 (2019)

44. Erukhimovich, I., de la Cruz, M.O.: Phase Equilibria and Charge Fractionation in Polydisperse Polyelectrolyte Solutions, pp. 1285–1290 (2004)
45. Zhang, D., Jiang, N., Chen, X., He, B.: Rheology of crosslinked entangled polymers: shear stiffening in oscillatory shear. J. Appl. Polym. Sci. **137**(9), 1–10 (2020)
46. Wang, S., Jiang, W., Jiang, W., Ye, F., Mao, Y., Xuan, S., et al.: Multifunctional polymer composite with excellent shear stiffening performance and magnetorheological effect. J. Mater. Chem. C. Mater. **2**(34), 7133–7140 (2014)
47. Liang, J., Zhang, X.H.: Rheological properties of SP in shock transmission application. J. Mater. Civ. Eng. **27**(9), 1–7 (2015)
48. Zhang, S., Wang, S., Wang, Y., Fan, X., Ding, L., Xuan, S., et al.: Conductive shear thickening gel/polyurethane sponge: a flexible human motion detection sensor with excellent safeguarding performance. Compos. Part A Appl. Sci. Manuf. **112**(May), 197–206 (2018)
49. Zhou, C., Wang, B., Zhang, F., Xu, K., Han, C., Hu, H., et al.: Micro-gels for impact protection. J. Appl. Polym. Sci. **130**(4), 2345–2351 (2013)
50. Seetapan, N., Fuongfuchat, A., Sirikittikul, D., Limparyoon, N.: Unimodal and bimodal networks of physically crosslinked polyborodimethylsiloxane: viscoelastic and equibiaxial extension behaviors. J. Polym. Res. **20**(7) (2013)
51. Zhang, F., Ju, P., Pan, M., Zhang, D., Huang, Y., Li, G., et al.: Self-healing mechanisms in smart protective coatings: a review. Corros. Sci. **144**(August), 74–88 (2018)
52. Wang, Y., Wang, S., Xu, C., Xuan, S., Jiang, W., Gong, X.: Dynamic behavior of magnetically responsive shear-stiffening gel under high strain rate. Compos. Sci. Technol. **127**, 169–176 (2016)
53. Zhang, S., Wang, S., Hu, T., Xuan, S., Jiang, H., Gong, X.: Study the safeguarding performance of shear thickening gel by the mechanoluminescence method. Compos. Part B Eng. **180**, 107564 (2020)
54. He, Q., Cao, S., Wang, Y., Xuan, S., Wang, P., Gong, X.: Impact resistance of shear thickening fluid/Kevlar composite treated with shear-stiffening gel. Compos. Part A Appl. Sci. Manuf. **106**, 82–90 (2018)
55. Tang, F., Dong, C., Yang, Z., Kang, Y., Huang, X., Li, M., et al.: Protective performance and dynamic behavior of composite body armor with shear stiffening gel as buffer material under ballistic impact. Compos. Sci. Technol. **2022**(218), 109190 (2021)
56. Liu, Z., Picken, S.J., Besseling, N.A.M.: Polyborosiloxanes (PBSs), synthetic kinetics, and characterization. Macromolecules. **47**(14), 4531 (2014)
57. Liu, X., Yu, K., Sun, J., Guo, W., Qian, K.: Effect of shear thickening gel on structure and properties of flexible polyurethane foam. Smart Mater. Struct. **30**(12), 125008 (2021)
58. Tu, H., Xu, P., Yang, Z., Tang, F., Dong, C., Chen, Y., et al.: Effect of shear thickening gel on microstructure and impact resistance of ethylene–vinyl acetate foam. Compos. Struct. **2023**(311), 116811 (2022)
59. Zhao, C., Xu, C., Cao, S., Xuan, S., Jiang, W., Gong, X.: Anti-impact behavior of a novel soft body armor based on shear thickening gel (STG) impregnated Kevlar fabrics. Smart Mater. Struct. **28**(7), 075036 (2019)
60. Wang, D.P., Zhao, Z.H., Li, C.H., Zuo, J.L.: An ultrafast self-healing polydimethylsiloxane elastomer with persistent sealing performance. Mater. Chem. Front. **3**(7), 1411–1421 (2019)
61. Wang, S., Xuan, S., Jiang, W., Jiang, W., Yan, L., Mao, Y., et al.: Rate-dependent and self-healing conductive shear stiffening nanocomposite: a novel safe-guarding material with force sensitivity. J. Mater. Chem. A. Mater. **3**(39), 19790–19799 (2015)
62. Boersma, W.H., Laven, J., Stein, H.N.: Shear thickening(dilatancy) in concentrated dispersions. AICHE J. **36**(3), 321–332 (1990)
63. Raghavan, S.R., Khan, S.A.: Shear-thickening response of fumed silica suspensions under steady and oscillatory shear. J. Colloid Interface Sci. **185**(1), 57–67 (1997)
64. Decker, M.J., Halbach, C.J., Nam, C.H., Wagner, N.J., Wetzel, E.D.: Stab resistance of shear thickening fluid (STF)-treated fabrics. Compos. Sci. Technol. **67**(3–4), 565–578 (2007)
65. Galindo-Rosales, F.J., Rubio-Hernández, F.J., Velázquez-Navarro, J.F.: Shear-thickening behavior of Aerosil® R816 nanoparticles suspensions in polar organic liquids. Rheol. Acta. **48**(6), 699–708 (2009)

66. Fischer, C., Bennani, A., Michaud, V., Jacquelin, E., Månson, J.A.E.: Structural damping of model sandwich structures using tailored shear thickening fluid compositions. Smart Mater. Struct. **19**(3), 035017 (2010)
67. Chen, Y., Xu, W., Zeng, G., Liu, W.: Shear-thickening behavior of precipitated calcium carbonate particles suspensions in glycerine. Appl. Rheol. **25**(1), 1–8 (2015)
68. Laha, A., Majumdar, A.: Interactive effects of p-aramid fabric structure and shear thickening fluid on impact resistance performance of soft armor materials. Mater. Des. **89**, 286–293 (2016)
69. Yang, W., Wu, Y., Pei, X., Zhou, F., Xue, Q.: Contribution of surface chemistry to the shear thickening of silica nanoparticle suspensions. Langmuir. **33**(4), 1037–1042 (2017)
70. Islam, E., Kaur, G., Bhattacharjee, D., Singh, S., Biswas, I., Verma, S.K.: Effect of cellulose beads on shear-thickening behavior in concentrated polymer dispersions. Colloid Polym. Sci. **296**(5), 883–893 (2018)
71. Son, H.S., Kim, K.H., Kim, J.H., Yoon, K.H., Lee, Y.S., Paik, H., jong.: High-performance shear thickening of polystyrene particles with poly(HEMA). Colloid Polym. Sci. **296**(9), 1591–1598 (2018)
72. Khodadadi, A., Liaghat, G., Vahid, S., Sabet, A.R., Hadavinia, H.: Ballistic performance of Kevlar fabric impregnated with nanosilica/PEG shear thickening fluid. Composites Part B. **162**(2018), 643–652 (2019)
73. Gürgen, S., Sofuoğlu, M.A.: Vibration attenuation of sandwich structures filled with shear thickening fluids. Compos. Part B Eng. **186**(January), 186 (2020)
74. Sahoo, S.K., Mishra, S., Islam, E., Nebhani, L.: Tuning shear thickening behavior via synthesis of organically modified silica to improve impact resistance of Kevlar fabric. Mater. Today Commun. **23**(August 2019), 100892 (2020)
75. Qin, J., Guo, B., Zhang, L., Wang, T., Zhang, G., Shi, X.: Soft armor materials constructed with Kevlar fabric and a novel shear thickening fluid. Compos. Part B Eng. **183**(November 2019), 107686 (2020)
76. Zhang, J., Wang, Y., Zhou, J., Zhao, C., Wu, Y., Liu, S., et al.: Intralayer interfacial sliding effect on the anti-impact performance of STF/Kevlar composite fabric. Compos. Part A Appl. Sci. Manuf. **145**(March), 106401 (2021)
77. Hao, J., Ding, J., Rutledge, G.C.: Shape-stable composites of electrospun nonwoven Mats and Shear-thickening fluids. ACS Appl. Mater. Interfaces. **14**(6), 8373–8383 (2022)
78. SampathKumar, D., Duraisamy, T., Pattabi, T., Mohankumar, A.K.: Static puncture resistance characteristics with various indenter nose shape geometry perforation of shear thickening fluid impregnated polypropylene fabric for soft Armour application. Polym. Polym. Compos. **30**, 1–10 (2022)
79. Liu, Z., Picken, S.J., Besseling, N.A.M.: Polyborosiloxanes (PBSs), synthetic kinetics, and characterization. Macromolecules. **47**(14), 4531–4537 (2014)
80. Li, C.S., Huang, X.C., Li, Y., Yang, N., Shen, Z., Fan, X.H.: Stab resistance of UHMWPE fiber composites impregnated with thermoplastics. Polym. Adv. Technol. **25**(9), 1014–1019 (2014)
81. Wang, S., Xuan, S., Wang, Y., Xu, C., Mao, Y., Liu, M., et al.: Stretchable polyurethane sponge scaffold strengthened shear stiffening polymer and its enhanced safeguarding performance. Appl. Mater. Interfaces. **8**, 4946–4954 (2016)
82. Wu, T., Chen, B.: Synthesis of multiwalled carbon nanotube-reinforced polyborosiloxane nanocomposites with mechanically adaptive and self-healing capabilities for flexible conductors. ACS Appl. Mater. Interfaces. **8**(36), 24071–24078 (2016)
83. Wang, D.P., Lai, J.C., Lai, H.Y., Mo, S.R., Zeng, K.Y., Li, C.H., et al.: Distinct mechanical and self-healing properties in two polydimethylsiloxane coordination polymers with fine-tuned bond strength. Inorg. Chem. **57**(6), 3232–3242 (2018)
84. Xiaoke, L., Kejing, Y., Qianqian, F., Kun, Q.: Shear thickening gel reinforced flexible polyurethane foam and its enhanced properties. Smart Mater. Struct. **28**(5), 055017 (2019)
85. Liu, L., Cai, M., Liu, X., Zhao, Z., Chen, W.: Ballistic impact performance of multi-phase STF-impregnated Kevlar fabrics in aero-engine containment. Thin-Walled Struct. **157**(29), 107103 (2020)

86. Sahoo, S.K., Mishra, S., Islam, E., Nebhani, L.: Tuning shear thickening behavior via synthesis of organically modified silica to improve impact resistance of Kevlar fabric. Mater. Today Commun. **23**, 23 (2020)
87. Zhao, C., Gong, X., Wang, S., Jiang, W., Xuan, S.: Shear Stiffening Gels for Intelligent Anti-Impact Applications, vol. 1, p. 100266. Cell Reports Physical Science (2020)
88. He, Q., Cao, S., Wang, Y., Xuan, S., Wang, P., Gong, X.: Impact resistance of shear thickening fluid/Kevlar composite treated with shear-stiffening gel. Compos. Part A Appl. Sci. Manuf. **106**, 82 (2018)
89. Wang, S., Xuan, S., Wang, Y., Xu, C., Mao, Y., Liu, M., et al.: Stretchable polyurethane sponge scaffold strengthened shear stiffening polymer and its enhanced safeguarding performance. ACS Appl. Mater. Interfaces. **8**(7), 4946 (2016)
90. Li, X., Zhang, D., Xiang, K., Huang, G.: Synthesis of polyborosiloxane and its reversible physical crosslinks. RSC Adv. **4**(62), 32894 (2014)
91. Zhang, S., Wang, S., Wang, Y., Fan, X., Ding, L., Xuan, S., et al.: Conductive shear thickening gel/polyurethane sponge: a flexible human motion detection sensor with excellent safeguarding performance. Compos. Part A Appl. Sci. Manuf. **112**, 197 (2018)
92. Palmer Richard Martin, Green PC. Energy Absorbing Material. Vol. 2, US Patent (2008)
93. Lee, Y.S., Kirkwood, K.M., Wagner, N.J., Egres, R.G., Lee, Y.S., Kirkwood, J.E., et al.: Liquid armor: protective fabrics utilising shear thickening fluids. researchgate.netRG Egres Jr, YS Lee, JE Kirkwood, KM Kirkwood, ED Wetzl, NJ WagnerProceedings of the 4th International Conference of Safety and, 2004•researchgate.net [Internet]. [cited 2023 Sep 23]; Available from: https://www.researchgate.net/profile/Yuan-Shin-Lee/publication/267546628_Liquid_armor_Protective_fabrics_utilizing_shear_thickening_fluids/links/54e75b900cf277664ffa2d5e/Liquid-armor-Protective-fabrics-utilizing-shear-thickening-fluids.pdf
94. Hasanzadeh, M., Mottaghitalab, V., Babaei, H., Rezaei, M.: The influence of carbon nanotubes on quasi-static puncture resistance and yarn pull-out behavior of shear-thickening fluids (STFs) impregnated woven fabrics. Compos. Part A Appl. Sci. Manuf. **88**, 263 (2016)
95. Woo, S.C., Kim, T.W.: High-strain-rate impact in Kevlar-woven composites and fracture analysis using acoustic emission. Compos. Part B Eng. **60**, 60 (2014)
96. Tan, Y., Ma, Y., Li, Y.: Warp-knitted spacer fabric flexible composites with high stability and low-velocity impact resistance infused by flock-reinforced shear thickening gel. Text. Res. J. **93**(17–18), 4246–4258 (2023)
97. Tu, H., Xu, P., Yang, Z., Tang, F., Dong, C., Chen, Y., et al.: Effect of shear thickening gel on microstructure and impact resistance of ethylene–vinyl acetate foam. Compos. Struct. **311**, 116811 (2023)
98. Tang, F., Dong, C., Yang, Z., Kang, Y., Huang, X., Li, M., et al.: Protective performance and dynamic behavior of composite body armor with shear stiffening gel as buffer material under ballistic impact. Compos. Sci. Technol. **218**, 109190 (2022)
99. Zhao, C., Wang, Y., Cao, S., Xuan, S., Jiang, W., Gong, X.: Conductive shear thickening gel/Kevlar wearable fabrics: a flexible body armor with mechano-electric coupling ballistic performance. Compos. Sci. Technol. **182**, 182 (2019)
100. Arora, S., Majumdar, A., Butola, B.S.: Structure induced effectiveness of shear thickening fluid for modulating impact resistance of UHMWPE fabrics. Compos. Struct. **210**, 41 (2019)
101. Kordani, N., Vanini, A.S.: Different method to make laminates by shear thickening fluid. Sci. Eng. Compos. Mater. **21**(3) (2014)
102. Deepak, S., Thirumalaikumarasamy, D., Thirumal, P., Dharun, T., Bhuvana, K.P., Joseph Bensingh, R., et al.: Preparation and characterization of shear thickening fluid coated polypropylene fabric for soft Armour application. J. Text. Inst. **112**(10), 1555 (2021)
103. Arora, S., Majumdar, A., Singh, B.B.: Interplay of fabric structure and shear thickening fluid impregnation in moderating the impact response of high-performance woven fabrics. J. Compos. Mater. **54**(28), 4387 (2020)
104. Srivastava, A., Majumdar, A., Butola, B.S.: Improving the impact resistance performance of Kevlar fabrics using silica based shear thickening fluid. Mater. Sci. Eng. A. **529**(1), 224 (2011)

105. Yeh, S.K., Zhuang, H.Y., Chen, Y.C., Tseng, P.C., Zheng, J.Y., Yang, L.Y., et al.: High-velocity impact performance of shear-thickening fluid/kevlar composites made by the padding process. Polym. Compos. **40**(8), 3040 (2019)
106. Majumdar, A., Butola, B.S., Srivastava, A., Bhattacharjee, D., Biswas, I., Laha, A., et al.: Improving the impact resistance of p-aramid fabrics by sequential impregnation with shear thickening fluid. Fibers Polym. **17**(2), 199 (2016)
107. Li, D., Wang, R., Guan, F., Zhu, Y., You, F.: Enhancement of the quasi-static stab resistance of Kevlar fabrics impregnated with shear thickening fluid. J. Mater. Res. Technol. **18**, 18 (2022)
108. Tan, Y., Li, Y., Ma, Y.: Evaluation of low-velocity impact responses of flexible protective materials with a novel, visible approach. Polym. Test. **113**, 107682 (2022)
109. Asija, N., Chouhan, H., Gebremeskel, S.A., Bhatnagar, N.: High strain rate characterization of shear thickening fluids using Split Hopkinson pressure Bar technique. Int. J. Impact Eng. **110**, 365 (2017)
110. Xie, Z., Chen, W., Liu, Y., Liu, L., Zhao, Z., Luo, G.: Design of the ballistic performance of shear thickening fluid (STF) impregnated Kevlar fabric via numerical simulation. Mater. Des. **226**, 111599 (2023)
111. Bilisik, K.: Two-dimensional (2D) fabrics and three-dimensional (3D) preforms for ballistic and stabbing protection: a review. Text. Res. J. **87**, 2275 (2017)
112. Zhang, X., Li, T.T., Peng, H.K., Lou, C.W., Lin, J.H.: Enhanced sandwich structure composite with shear thickening fluid and thermoplastic polyurethanes for high-performance stab resistance. Compos. Struct. **280**, 114930 (2022)
113. Sun, L.L., Xiong, D.S., Xu, C.Y.: Application of shear thickening fluid in ultra high molecular weight polyethylene fabric. J. Appl. Polym. Sci. **129**(4), 1922 (2013)
114. Li, T.T., Cen, X., Peng, H., Ren, H., Han, L., Lou, C.W., et al.: Rheological response and quasi-static stab resistance of STF/MWCNTs-impregnated aramid fabrics with different textures. J. Ind. Text. **50**(3), 380 (2020)
115. Feng, X., Li, S., Wang, Y., Wang, Y., Liu, J.: Effects of different silica particles on quasi-static stab resistant properties of fabrics impregnated with shear thickening fluids. Mater. Des. **64**, 456 (2014)
116. Ismail, M.F., Sultan, M.T.H., Hamdan, A., Shah, A.U.M., Jawaid, M.: Low velocity impact behaviour and post-impact characteristics of kenaf/glass hybrid composites with various weight ratios. J. Mater. Res. Technol. **8**(3), 2662 (2019)
117. Asija, N., Chouhan, H., Amare Gebremeskel, S., Bhatnagar, N.: Impact response of shear thickening fluid (STF) treated ultra high molecular weight poly ethylene composites – study of the effect of STF treatment method. Thin-Walled Struct. **126**, 16 (2018)
118. Zhang, Q., Liu, Z., Qin, Z., Yan, R., Jia, L.: Characterization of macroscopic impact-resistant behavior of shear thickening fluid impregnated ultra-high molecular weight polyethylene fiber flexible composites. Compos. Commun. **25**, 25 (2021)
119. Cheng, J., Ye, L., Fu, K., Wang, H.: Effect of striker shape on impact energy absorption of a shear thickening fluid. Compos. Commun. **23**, 23 (2021)
120. Kim, Y.H., Sathish Kumar, S.K., Park, Y., Kwon, H., Kim, C.G.: High-velocity impact onto a high-frictional fabric treated with adhesive spray coating and shear thickening fluid impregnation. Compos. Part B Eng. **185**, 185 (2020)
121. Tan, Y., Ma, Y., Li, Y.: Shear thickening fabric composites for impact protection: a review. Text. Res. J. **93**, 1419 (2023)
122. Fan, X., Liu, S., Wang, B., Wang, Y., Wang, S., Gong, X.: An experimental and numerical study of sutural composites with shear stiffening gel core. J. Mater. Sci. **57**(2), 1328–1344 (2022)
123. Ying, P., Shen, W., Xia, Y.: Shear stiffening gel under low-speed impact: mechanical behavior characterization and protective performance analysis. Mater. Des. **231**, 112047 (2023)

Index

S. Gürgen (ed.), *Smart Systems with Shear Thickening Fluid*,
https://doi.org/10.1007/978-3-031-53570-3

SPRINGER NATURE

GPSR Compliance

The European Union's (EU) General Product Safety Regulation (GPSR) is a set of rules that requires consumer products to be safe and our obligations to ensure this.

If you have any concerns about our products, you can contact us on ProductSafety@springernature.com

In case Publisher is established outside the EU, the EU authorized representative is:

Springer Nature Customer Service Center GmbH
Europaplatz 3
69115 Heidelberg, Germany

The manufacturer's authorised representative in the EU is Springer Nature Customer Service Centre GmbH, Europaplatz 3, 69115 Heidelberg, Germany. If you have any concerns regarding our products, please contact ProductSafety@springernature.com

Printed and bound by CPI Group (UK) Ltd, Croydon, CR0 4YY
01/10/2026
02225916-0003